Oleg Pykhteev

Characterization of Acoustic Waves in Multi-Layered Structures

Oleg Pykhteev

Characterization of Acoustic Waves in Multi-Layered Structures

Mathematical modeling of acoustic waves propagating in piezoelectric and elastic multi-layered structures

Südwestdeutscher Verlag für Hochschulschriften

Imprint
Any brand names and product names mentioned in this book are subject to trademark, brand or patent protection and are trademarks or registered trademarks of their respective holders. The use of brand names, product names, common names, trade names, product descriptions etc. even without a particular marking in this work is in no way to be construed to mean that such names may be regarded as unrestricted in respect of trademark and brand protection legislation and could thus be used by anyone.

Publisher:
Südwestdeutscher Verlag für Hochschulschriften
is a trademark of
Dodo Books Indian Ocean Ltd., member of the OmniScriptum S.R.L Publishing group
str. A.Russo 15, of. 61, Chisinau-2068, Republic of Moldova Europe
Printed at: see last page
ISBN: 978-3-8381-2518-3

Zugl. / Approved by: München, TU, Diss., 2010

Abstract

The work is devoted to the modeling of acoustic waves in multi-layered structures surrounded by a fluid and consisting of different kinds of materials including piezoelectric materials and composite multilayers. It consists of three parts. The first part describes the modeling of an acoustic sensor by the finite element method. The existence and uniqueness of a time-harmonic solution are rigorously established under physically appropriate assumptions. It is shown that the elimination of nonzero eigenvalues of the arising Helmholtz-like system can be achieved by introducing an additional term that causes the damping effect near the boundary. We also establish the well-posedness of the discretized problem and the convergence of Ritz-Galerkin solutions to the solution of the exact problem. Besides, we derive a domain decomposition schema for the numerical treatment of the problem. Finally, the results of 3D simulations are presented.

The second part of the work presents a semi-analytical method for the fast characterization of plane acoustic waves in multi-layered structures. The method identifies plane waves that can possibly exist in a given structure, determines their velocities, and computes the dispersion relation curves. It handles multi-layered structures composed of an arbitrary number of layers made of different material types. The influence of a surrounding fluid or dielectric medium can also be taken into account. The software implementing this approach is presented.

The third part investigates a number of issues of the homogenization theory for linear systems of elasticity. The results presented here are exploited in the previous part for the modeling of acoustic waves in composite materials called multilayers. Such materials consist of huge number of very thin periodic alternating sublayers. In this part we rigorously derive the limiting equations in the general three-dimensional case by the two-scale method and establish an error estimate for the case where the right-hand side is in L^2. The homogenization of laminated structures is considered as a special case. For this case an explicit formula for the elasticity tensor of the homogenized material is derived.

Acknowledgements

First of all I would like to thank Nikolai Botkin, for a lot of time and patience he spent in our numerous discussions, for his help and continuous encouragment during the work. Without his influence this work would not have appeared.

I am deeply thankful to my supervisor Karl-Heinz Hoffmann, for giving me the opportunity to work at TUM, for his support over the years, and his valuable advices.

Contents

List of Figures

List of Tables

1 Introduction

1.1 Motivation and Object

Acoustic Wave devices have been in industrial applications for many decades and are becoming more and more popular. Traditionally consumed mostly in telecommunications they now conquer new increasingly growing application areas in the auto industry, metallurgy, medicine, and domestic appliances.

Exploiting surface acoustic shear waves gives rise to the development of tiny sensors with very high mass sensitivity that are especially well suitable for detecting chemicals in liquids. Such sensors are being widely utilized in medical and technical applications.

The development of such high sensitive sensors is hardly imaginable without preliminary mathematical modeling aimed at the optimization of the layout and size of structural elements, identification and detailed study of relevant physical processes, and estimation of the sensitivity and performance limits.

The work was initially motivated by the development of a biosensor at the research center caesar. This biosensor serves for the detection and quantitative measurement of microscopic amounts of biological substances. The underlying operating principle is based on the generation and detection of horizontally polarized surface acoustic waves (SAW) in a piezoelectric substrate. The substrate is a cut of a piezoelectric crystal oriented in such a way that the excited wave is horizontally polarized. This wave is guided by an elastic layer welded on the top of the substrate so that one can speak about Love waves (see [47]). Mechanical displacements in such waves are free of the transversal component, and, therefore, no appreciable energy is radiated into the contacting liquid, which makes shear Love waves perfectly suitable for liquid sensing applications. In this area Love wave sensors have the highest sensitivity in comparison with all the others acoustic sensors (see [23]).

During the work on the biosensor we have developed a modeling approach that is applicable to a wide range of multi-layered structures comprising those occurring in the specific biosensor under consideration. Moreover, our methods can easily be extended to cover other types of media constituting the layers. In particular, many physicists are interested in the characterization of acoustic waves propagating in composite materials with very thin alternating sublayers. A large number of the sublayers and their small thickness in comparison with the wavelength makes any direct modeling of such materials impossible. So that homogenization theory for elastic materials has to be involved.

1.2 State of the Art

There exists an abundant literature on characterization of acoustic waves in crystals. Among others, surface acoustic waves are under particular intensive study. The fundamental work [40] describes the connection between elasticity theory and wave propagation including surface waves of different types. The dispersion relations are derived for two-layered structures composed of isotropic materials. The monograph [4] contains a comprehensive description of elasticity theory for crystals. Mathematical methods for the study of reflection and scattering of acoustic Rayleigh waves are presented. The book [75] presents the linear elasticity theory for piezoelectric media and establishes relations between mathematical models and engineering representations of characteristics of piezoelectric materials. The monograph [60] treats principles of mathematical modeling of wave propagation. This includes elasticity and piezoelectric equations for crystal media, boundary conditions, interface requirements, and dispersion relations for various wave types. The excitation of acoustic waves in piezoelectric crystals with embedded electrodes as well as wave guides and guiding layers are considered from the mathematical and technical points of view.

The works [14], [39], [15], [31], and [32] study shear wave sensors regarding measurements in liquids. The mathematical model developed in [14] is based on harmonic analysis. It is assumed that the sensor is infinite in the horizontal directions. The presence of a liquid is taken into account by an additional viscoelastic term in the wave equation. In [39], the effect of the viscosity of the liquid on the noise of the output signal is analyzed for the case of a small viscosity. The investigation is based on formulas from [4]. In [15], the dependence of the sensitivity on the thickness of the guiding layer is studied using formulas from [4]. Numerical results are in a good agreement with laboratory tests as long as the thickness of the guiding layer does not exceed a certain value. For thick guiding layers, large deviations from measured values arise. The same model is examined in [31] using Fourier analysis. An effective way to analyze the dependence of the sensitivity on the liquid viscosity is proposed in [32].

Another approach to analyze surface acoustic waves in multi-layered structures and semi-infinite substrates is based on using orthogonal functions [13]. In particular, Laguerre [34, 35] and Legendre polynomials [43] are of most use. The work [42] studies conceptual advantages and limitations of the Laguerre polynomial approach. Among other things it is shown there that Laguerre polynomial method cannot be used to study leaky surface acoustic waves.

Much literature exists on investigation of acoustic waves in multi-layered structures by the so-called transfer matrix method. A good overview on this subject can be found in [49]. The method introduces transfer matrices that describe the displacements and stresses at the bottom of the layer with respect to those at the top of the layer. The matrices for all the layers are then coupled by interface conditions to yield a system matrix for the complete system. Thus the displacements and stresses at the bottom of the multi-layered structure are related to those at the top of the structure. Modal or response solutions could then be found by application of the appropriate boundary conditions. The method, first proposed in [70], has been pursued and enhanced by many authors. The works [66, 58, 19, 65, 73] extend the original theory to leaky waves by introducing a real exponential factor in the wave equation. This is achieved by allowing either the frequency or the wavenumber to be generally complex. A number of works are devoted to the treatment of instabilities that arise when layers of a relatively large thickness are present and high-frequency are considered. One approach is based on a rearranging the equations in such a way that they do not become ill-conditioned [16, 71, 1, 45, 9]. Another approach is to employ a "Global matrix" in which a large single matrix is assembled, consisting of all of the equations for all of the layers [36, 64, 62, 63, 48]. The work [57] presents a program for computing dispersion relations in multi-layered structures based on the latter approach. Though the numerical algorithm has a number of similarities with the algorithm proposed in Chapter 3 of this work, it has limitations when dealing with anisotropic materials and does not cover piezoelectric media. The last is apparently a consequence of aiming at applications in nondestructive testing. The transfer matrix method has been initially applied in geology. Hence it has been focused on pure elastic structures only. Meanwhile, some extensions to piezoelectric materials have appeared [72, 7].

The works cited above are mostly based on harmonic analysis. When applied to the simulation of real devices, these techniques require significantly simplifying assumptions. This limits the practical use of the models proposed. Many quantitative questions remain unanswered, especially, if the object under consideration has a complicated geometrical structure, e.g., if it consists of several layers with extremely different thicknesses or embraces other obstacles like electrodes. Moreover, in order to avoid the consideration of wave reflections on the faces, it is often assumed that the layers are infinite in lateral dimensions. Therefore, important effects such as the excitation of parasite frequencies caused by wave reflections cannot be seized. The influence of internal obstacles, e.g., electrodes is not taken into account as well.

Modeling with finite elements or volumes is more promising, because it allows to take into account and quantitatively estimate the above-mentioned effects and nonlinear interactions with the surrounding liquid. One of the difficulties here is the modeling of the liquid-solid interaction. J. L. Lions [46] treats this problem by means of a variable transformation, which yields some non local in time but well-posed integro-differential equations. Another method for the treatment of the contact between a solid body and a liquid consist in a suitable penalization of the interface conditions. This reduces the original problem to a control problem [24].

Due to advances in computer performance the finite element approach to modeling of acoustic devices became more popular during the last two decades [44, 17, 52, 59, 27, 20, 26]. However, most of the works lack the rigorous mathematical analysis of models utilized and do not prove the convergence of numerical methods. Many works use reduced two-dimensional models or do not consider the contacting liquid.

A theoretical investigation of linearized equations of piezoelectricity and their treatment by the finite element method is given in [21] and [22]. Solvability conditions for the time-harmonic case are analyzed. The original system of equations is transformed to the associated Schur complement system for which a modification of the Fredholm alternative is proved. The works [37, 38] extend the model developed in [21] and [22] by considering the acoustic streaming in fluid-filled microchannels located on the top surface of a piezoelectric.

The fundamental work on homogenization of linear systems of elasticity is due to Oleinik, Shamaev, and Yosifian [56]. The book contains a lot of theoretical results. In particular, it presents the limiting equations (though without derivation) and establishes an error estimate for the case where the right-hand side is in H^1. The rigorous derivation of the limiting equation by Tartar's method of oscillating test functions (see [68, 69]) can be found in [11]. The book [33] can be referred as a comprehensive monograph on homogenization theory of partial differential equations. The homogenization of elasticity tensors is based here on the theory of G-convergence. Among many other things the book also derives an explicit formula for the homogenized tensors in the case of layered materials. However, the derivation is given for isotropic materials only.

1.3 Overview

The thesis presents two approaches to modeling of acoustic waves in multi-layered structures.

The first part of the work is devoted to the modeling of the acoustic sensor mentioned in Section 1.1 by the finite element method. The main advantage of this method is the ability to take into account the exact parameters of the sensor such as the shape of the electrodes, their position, electroconductivity properties. This allows to estimate important characteristics of the biosensor and effects caused by the scattering of waves. This approach is described in details in Chapter 2.

The FE-model is developed under the following assumptions:

1. We consider linear material laws for solids and neglect nonlinear terms in the description of the fluid. This is reasonable because the displacements and velocities are very small for the structure under consideration.

2. Only time-periodic solutions are considered.

3. The damping effect at the sides of the biosensor is modeled by an additional term in the governing equations. This term is zero inside the structure and grows linearly in some damping area as it approaches the boundaries.

4. We introduce a small term describing the dielectric dissipation in the piezoelectric substrate (see Section 2.4). Due to its smallness, the term has no significant influence on the result, but it plays an important role in the proof of the well-posedness of the model while preserving the physical meaning.

5. The liquid-solid interface is treated by means of the variable transformation as described in [46].

The presented model is an extension and modification of the model described in [6]. Among other things, the extended model accounts for special bristle-like layers arising in some applications at the liquid-solid interface. The simulation of them is based on the homogenization technique developed in [28]. We also provide a rigorous mathematical analysis of the model. Beside the proof of the well-posedness the chapter investigates numerical issues. It establishes the convergence of the Ritz-Galerkin solutions to the exact one and proposes a numerical scheme based on domain decomposition. Finally, the results of 3D-simulations are presented.

Though the FE-approach provides accurate results, it has a number of disadvantages. The main disadvantage is the laboriousness of the computer implementation. The computations are very high time- and resource-consuming due to a very small wavelength. In order to resolve the wave structure appropriately, a large number of elements in the wave traveling direction is required. The number of degrees of freedom lies in the range of $10^7 - 10^8$, which makes simulations on stand-alone ordinary computers impossible and requires parallel computing. Another disadvantage is the inflexibility when optimizing the constructive features of the sensor. Changes in the geometry, adding or removing layers involve essential changes in the FE-discretization and require the complete recalculation.

These disadvantages suggested us to look for lighter-weighted and more flexible approaches to the modeling, which would allow to obtain preliminary results faster for the price of a relaxed mathematical model. The approach described in Chapter 3 is based on the harmonic analysis of plane waves propagating in multi-layered structures unbounded and homogeneous in the horizontal directions. This method allows to identify traveling waves feasible in a given structure and derive the corresponding dispersion relations, i.e. the relations between the propagation velocity and the wave frequency. As a rule, the analytical derivation of dispersion relations in multi-layered structures is not realizable. Therefore the method described in this work is semi-analytical and essentially relies on numerical procedures.

The assumptions of the unboundedness and homogeneity of the structure in the horizontal directions prevent this method from the accurate simulation of real devices. The method is not able to take into account a number of important parameters relevant to sensors, such as the dimensions, the shape and layout of electrodes. On the other hand, it provides very important preliminary information such as the wavelength and the displacement profile in the transversal direction including the attenuation rates of waves in the substrate and in the fluid. Such information is very important for choosing optimal reliable finite element approximations. Besides, the assumption of the unboundedness is quite relevant for the characterization of propagating acoustic waves, because real sensor chips are usually embedded up to the surface in some viscose damping medium to exclude the reflection of waves on the side and bottom faces. To some extent this is equivalent to the above mentioned acoustic unboundedness.

As this method was initially applied to the modeling of the biosensor, it became clear that the algorithm lying in the base of it can be used for the characterization of acoustic waves in a much wider range of structures than that of the biosensor. Namely, the method can easily be adjusted to almost arbitrary multi-layered structure consisting of any finite number of layers and can be applied for the characterization of any type of plane waves, not only surface acoustic waves. Exploiting this idea we developed a computer program which calculates dispersion relations in arbitrary structures specified by the user. The program has a user-friendly interface that allows to manipulate with layers and materials in a simple way.

Initially, the possible types of materials were limited to piezoelectric and isotropic elastic materials, the surrounding medium could be absent or be a weak-compressive fluid. Later on, reacting on the needs of simulations the list of accessible materials and media was significantly extended. The careful examination of the electric field in the structure forced us to enrich isotropic materials with electric properties extending the corresponding mathematical treatment. For the same reason a dielectric surrounding media (like gas or vacuum) was introduced. Significant efforts were put to the modeling of thin bristle-like layers contacting with a fluid. In order to handle such layers we exploit the homogenization technique described in [28] which enables us to reduce the problem to the case of a bulk layer. Such layers were as well successfully integrated into the program, which involved the entire numerical implementation of the homogenization procedures.

The necessity in another kind of homogenization arises when dealing with composite materials consisting of a large number of periodically alternating thin sublayers. A typical example of such materials are so called multilayers (see for example [25]). The direct modeling of them is hardly possible due to a large number of sublayers and their small thickness in comparison to the wavelength. Therefore the original composite materials are replaced with an averaged one whose properties are derived as the thickness of the repeating set of the sublayers goes to 0 and their number goes to infinity. This involves the homogenization theory for linear systems of elasticity. This topic is the subject of Chapter 4. In this chapter we rigorously derive the limiting equations in general three-dimensional case by the two-scale method and establish an error estimate for the case where the right-hand side is in L^2. The homogenization of laminated structures is of particular interest and considered as a special case. For this case an explicit formula for the elasticity tensor of the homogenized material is derived. This enabled us to extend the presented program for calculating acoustic waves with this kind of composite materials.

The main practical result of this work is the developed computer program that represents a powerful modeling tool for the fast characterization of acoustic waves in multi-layered structures. A wide range of the supported material types and the ability to simulate different types of waves make this tool applicable in many application domains including geophysics, non-destructive testing, and design of acoustic devices. The ability to simulate piezoelectric materials, bristle-like layers, and surrounding fluids makes the program especially useful for engineers working on acoustic sensors. The program can be very helpful at early stages of designing acoustic devices because it allows to obtain many important wave parameters very quickly. For example, one can quickly estimate the sensitivity of a sensor depending on many construction parameters such as the thickness of layers, their mechanical and electrical properties, the properties of the surrounding medium.

More accurate but time-expensive simulations of acoustic sensors can be done by the finite element method. The presented final element model of the biosensor can be applied to a wide range of similar acoustic based devices. The crucial assumption for the well-posedness of the model is the presence of damping area surrounding the device. As long as this condition remains the developed theory is applicable.

Finally, the both approaches provide the most efficient way to simulate SAW devices when applied together. The preliminary results obtained by the method based on dispersion relations can then be used to adjust and optimize the finite element model. On the other hand, they can be used for fast verification of results of FE-simulations.

Another result that can be useful for physicists and engineers is the rigorously derived explicit formula for the calculation of the elasticity tensor of multilayers. The elastic properties of multilayers are of extreme importance. There are many works devoted to the measurement and calculation of Young's modulus for such materials. The derived formula can be very helpful for people working in this area.

2 Finite Element Model of Acoustic Biosensor

2.1 Introduction

This chapter is devoted to the simulation of the biosensor mentioned in Section 1.1 by the final element method. The biosensor serves for the detection and quantitative measurement of a specific protein in a contacting liquid. The key role in the detection process is played by the so-called *aptamers*. Aptamers are special molecules that bind to a specific target protein selectively. They are designed based on the protein to detect.

As depicted in Figure 2.1 the biosensor consists of several layers. The bottom layer is a substrate made of a piezoelectric material. Two groups of electrodes are deposited on top of it. An acoustic wave is excited in the substrate by applying alternate voltage to the input electrodes. It travels then through the whole structure towards the output electrodes that serve to identify its characteristics at the end of the path. The surface of the top layer contacts with the liquid. It is covered with the aptamer receptors. If the target protein is present in the liquid, it gets caught by the aptamers so that the mass of the whole structure increases and the wave travels slower. The arising phase shift at the end of the path is then identified by the output electrodes.

We state a three-dimensional mathematical model that describes the biosensor structure consisting of the following five layers (see Figures 2.1 and 2.2): a piezoelectric substrate made of α-quarz, a guiding layer made of silicon dioxide, a gold shilding layer, a bristle-like aptamers layer and a liquid layer considered as a weakly compressible viscous fluid. The biosensor is embedded up to the surface into a very viscous damping medium to exclude the reflection of waves on side and bottom faces. The full coupling between the deformation and electric fields is assumed.

The chapter is organized as follows. Section 2.3 states the governing equations, the boundary and interface conditions. Section 2.4 describes the derivation of the weak formu-

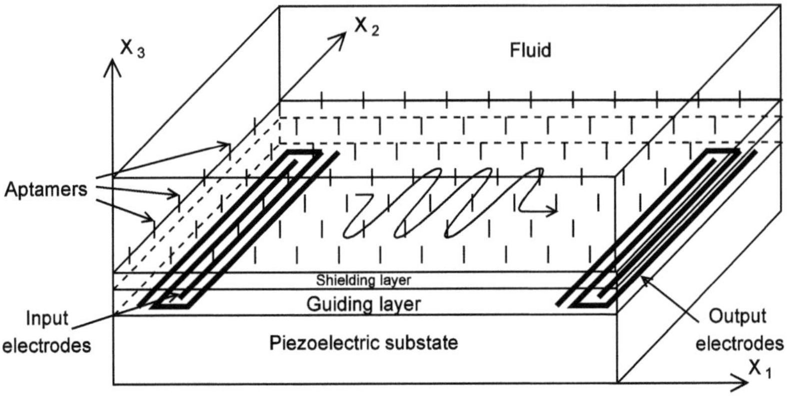

Figure 2.1: Sketch of the biosensor.

lation of the problem and provides an analysis of its properties. The well-posedness of the model is then established in Section 2.5. Section 2.6 is devoted to the numerical treatment of the model. It shows the well-posedness of the discrete problem and the convergence of the discrete solution to the solution of the original problem. Besides, it describes the numerical treatment by the domain decomposition approach. Finally, Section 2.7 presents the results of the finite element simulations.

2.2 Notation

We use the cartesian coordinate system (x_1, x_2, x_3). The x_3-axis is orthogonal to the sensor surface; the x_1-axis is parallel to the wave propagation direction.

Let u_1, u_2, and u_3 be the displacements in the x_1, x_2, and x_3 directions, respectively; v_1, v_2, and v_3 the velocity components; p the pressure; ϱ the density.

Vectors are distinguished from scalar quantities by writing the quantity in a bold font. For example,

$$\boldsymbol{u} = (u_1, u_2, u_3)^T$$

is the displacement vector, and

$$\boldsymbol{v} = (v_1, v_2, v_3)^T$$

is the velocity vector.

The Einstein's summation convention is exploited throughout the work. The subscript t when applied to a function denotes the derivative with respect to time.

Denote by $\varepsilon(\boldsymbol{w})$ the symmetric part of the gradient of a vector function $\boldsymbol{w}$, i.e.

$$\varepsilon_{ij}(\boldsymbol{w}) := \frac{1}{2}\left(\frac{\partial w_i}{\partial x_j} + \frac{\partial w_j}{\partial x_i}\right).$$

The symmetric part of the gradient of the displacement vector is denoted by ε, i.e.

$$\varepsilon := \varepsilon(\boldsymbol{u}).$$

Note that ε is the usual infinitisimal strain tensor.

The symmetric part of an arbitraty second-rank tensor ξ is denoted by $\operatorname{sym}(\xi)$.

To distinguish functions and parameters related to different media we introduce the following sub- and superscripts that indicate the medium:

- f - fluid,
- a - aptamer layer,
- s - shielding layer (usually made of gold),
- g - guiding layer (usually made of $\mathtt{SiO_2}$),
- p - piezoelectric substrate.

Open domains occupied by media are denoted by Ω with the corresponding subscript. By Ω_p^d, Ω_g^d and Ω_s^d denote the damping subdomains (see below) of Ω_p, Ω_g and Ω_s respectively. Neighboring domains with the interface between them are indicated by the combination of the corresponding subscripts. For example, $\Omega_{pgs} = \operatorname{int}(\overline{\Omega_p \cup \Omega_g \cup \Omega_s})$. The domain occupied by the whole device is denoted by Ω. By definition, $\Omega = \Omega_{pgsaf}$.

We use the letter C to represent a generic positive constant that may take different values at different occurrences.

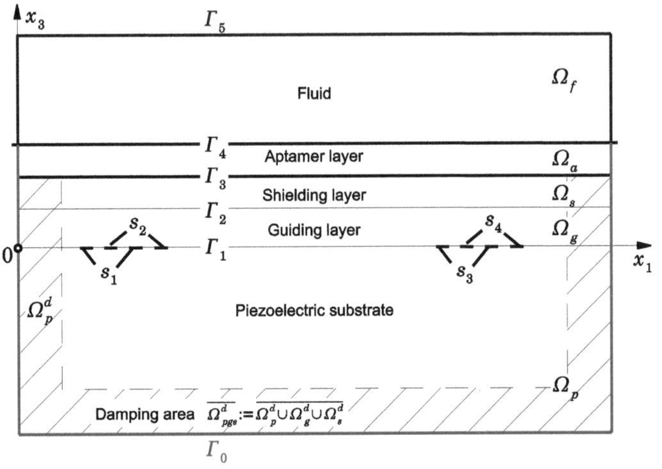

Figure 2.2: Cross section of the biosensor.

2.3 Governing equations and Conditions

2.3.1 Governing Equations

We consider linear material laws for solids (see [75]) and neglect nonlinear terms in the description of the fluid. This is reasonable because the displacements and velocities are very small for the structure under consideration.

The electrodes lying on the substrate are very thin. Their thickness is in the range of 200 to 300 nm. This enables us to simplify the geometry of the structure by assuming the electrodes to be plain. This simplification implies that the two-dimensional domain occupied by the electrodes is a part of the plain interface between the substrate and the guiding layer. We denote this domain by $S \subset \mathbb{R}^2$. Two plain domains occupied by two alternating groups of the input electrodes are denoted by S_1 and S_2. The domains of the output electrodes are indicated by S_3 and S_4. By definition $S = S_1 \cup S_2 \cup S_3 \cup S_4$.

We also neglect the mechanical influence of the electrodes. This influence is insignificant because the electrodes are very thin and narrow, and therefore their mass is tiny. Thus we have no equation for them. However, their size, shape and layout determine the geometry of the electrical boundary conditions. Hence they are still taken into account by the model.

The damping effect of the surrounding viscous medium is simulated by introducing an additional viscous term in the governing equations. This term is zero outside some damping domain and grows linearly as it approaches the boundaries (see Figure 2.2).

We state now the governing equations for all the layers.

Piezoelectric Substrate.
The constitutive relations for the piezoelectric substrate in the case of small deformations are of the form

$$\sigma_{ij} = G_{ijkl}\varepsilon_{kl} - e_{kij}E_k, \tag{2.1}$$

$$D_i = \epsilon_{ij}E_j + e_{ikl}\varepsilon_{kl}. \tag{2.2}$$

Here, σ_{ij} and ε_{kl} are the stress and the strain tensors, $\boldsymbol{D}$ and $\boldsymbol{E}$ denote the electric displacement and the electric field; ϵ_{kl}, e_{kij}, and G_{ijkl} denote the material dielectric tensor, the stress piezoelectric tensor, and the elastic stiffness tensor, respectively. The momentum conservation law and Gauss's law yield the following governing equations:

$$\begin{aligned} \varrho\boldsymbol{u}_{tt} - \mathrm{div}\sigma - \mathrm{div}(\beta(x)\nabla\boldsymbol{u}_t) = 0,\\ \mathrm{div}\boldsymbol{D} = 0. \end{aligned} \tag{2.3}$$

Here the term with $\beta(x)$ expresses the damping on the side boundaries of the device. The function $\beta(x)$ is assumed to be zero outside of the damping region $\Omega_p^d \cup \Omega_g^d \cup \Omega_s^d$ and it grows linearly up to some $\beta_0 > 0$ towards the side boundaries of the sensor. Substituting (2.1) and (2.2) into (2.3) yields the following governing equations for the displacements and the electric potential in the substrate:

$$\begin{cases} \varrho u_{i\,tt} - G_{ijkl}\dfrac{\partial^2 u_l}{\partial x_j \partial x_k} - e_{kij}\dfrac{\partial^2 \varphi}{\partial x_k \partial x_j} - \mathrm{div}(\beta(x)\nabla u_{i\,t}) = 0, \quad i = 1,2,3, \\ -\epsilon_{ij}\dfrac{\partial^2 \varphi}{\partial x_i \partial x_j} + e_{ikl}\dfrac{\partial^2 u_l}{\partial x_i \partial x_k} = 0 \end{cases} \quad \text{in } \Omega_p, \tag{2.4}$$

where φ is the electric potential, i.e.

$$\boldsymbol{E} = -\nabla\varphi.$$

Later on we will use the following important properties of the tensors G, e and ϵ:

- *Symmetry*

$$G_{ijkl} = G_{klij} = G_{jikl}, \quad \epsilon_{ij} = \epsilon_{ji}, \quad e_{ikl} = e_{ilk}, \quad i, j, k, l = 1, 2, 3.$$

- *Positiveness*

$$\epsilon_{ij} v_i v_j \geqslant C|\boldsymbol{v}|^2, \tag{2.5}$$

$$G_{ijkl}\xi_{ij}\xi_{kl} \geqslant C(\xi : \xi), \tag{2.6}$$

for all $\boldsymbol{v} \in \mathbb{R}^3$, all second-rand symmetric tensors ξ and some positive constant C.

Schielding and guiding layer.
The schielding layer is conductive so that there is no electric field inside of it. The guiding layer is an insultor, but it is very thin and the electric field at the upper surface is zero because it contacts to the schielding layer (see Figure 2.2). For this reason we consider the electric field in the whole guiding layer to be neglible. This implies that the stress has no electrically originated component. Furthermore, both materials are assumed to be isotropic. The stress tensor is then of the form

$$\sigma_{ij} = \lambda \delta_{ij} \varepsilon_{kk} + 2\mu \varepsilon_{ij},$$

where λ and μ are Lamé parameters. The corresponding governing equation is then

$$\varrho \boldsymbol{u}_{tt} - \mu \Delta \boldsymbol{u} - (\lambda + \mu)\nabla(\mathrm{div}\boldsymbol{u}) - \mathrm{div}(\beta(x)\nabla \boldsymbol{u}_t) = 0 \qquad \text{in } \Omega_g \cup \Omega_s. \tag{2.7}$$

Fluid.
In the fluid layer, the Navier-Stokes equation and mass conservation equation hold:

$$\begin{aligned} \varrho\,(\boldsymbol{v}_t + (\boldsymbol{v}\cdot\nabla)\boldsymbol{v}) &= -\,\nabla p + \nu\Delta\boldsymbol{v} + (\zeta + \frac{\nu}{3})\nabla(\mathrm{div}\,\boldsymbol{v}), \\ \varrho_t &= -\,\mathrm{div}(\varrho\boldsymbol{v}), \end{aligned} \tag{2.8}$$

where ν and ζ are the dynamic and volume viscosities of the fluid, respectively. We exploit now the fact that the fluid is weakly compressible and assume the following relation between the density and pressure (see [41]):

$$\varrho(p) = \varrho_0 + \left.\frac{\partial \varrho}{\partial p}\right|_{\varepsilon} (p - p_0). \tag{2.9}$$

Here $\left.\frac{\partial \varrho}{\partial p}\right|_{\varepsilon}$ means the density change under a constant entropy. It is assumed to be constant. ϱ_0 and p_0 are the constant static density and pressure respectively. Futhermore, due to the weak compressibility the changes in pressure and density are small, i.e. $|p - p_0| \ll p_0$ and $|\varrho - \varrho_0| \ll \varrho_0$. We rewrite now the system (2.8) neglecting all the terms of the second order of smallness. Besides, we assume ν, ζ, and variations in div $\boldsymbol{v}$ are small so that the term $(\zeta + \nu/3)\nabla(\operatorname{div} \boldsymbol{v})$ can also be neglected. This yields the following governing equations for the fluid:

$$\begin{cases} \varrho_0 \dfrac{\partial \boldsymbol{v}}{\partial t} + \nabla p - \nu \Delta \boldsymbol{v} = 0, \\ \gamma p_t + \operatorname{div} \boldsymbol{v} = 0 \end{cases} \quad \text{in } \Omega_f, \tag{2.10}$$

where $\gamma := \left.\frac{1}{\varrho_0}\frac{\partial \varrho}{\partial p}\right|_{\varepsilon}$ is the compressibility of the fluid. The corresponding expression for the stress is

$$\sigma_{ij} = -p\delta_{ij} + \nu \frac{\partial v_i}{\partial x_j},$$

where δ_{ij} is the Kronecker delta.

Aptamer Layer.

In order to treat the aptamer structure at the liquid-solid interface we apply the homogenization technique developed in [28]. The original bristle-like structure surrounded by the fluid is replaced by an averaged material whose properties are derived as the number of bristles goes to infinity whereas their thickness goes to zero. The height remains constant. We end up with a new layer with thickness equaled to the height of the aptamers. The governing equation for this layer reads (see [28]):

$$\varrho u_{i\,tt} - \hat{G}_{ijkl} \frac{\partial^2 u_l}{\partial x_j \partial x_k} - \hat{P}_{ijkl} \frac{\partial^2 u_{l\,t}}{\partial x_j \partial x_k} = 0 \quad \text{in } \Omega_a. \tag{2.11}$$

The stress tensor of the homogenized material is of the form

$$\sigma_{ij} = \hat{G}_{ijkl} \frac{\partial u_l}{\partial x_k} + \hat{P}_{ijkl} \frac{\partial u_{l\,t}}{\partial x_k}.$$

Here the term containing the tensor $\hat{P}$ describes the viscous damping that originates from the fluid part of the bristle structure. The term with $\hat{G}$ represents elastic stresses. The density ϱ here is determined by the density of the fluid and the density of the aptamers. We will need the following properties of $\hat{G}$ and $\hat{P}$:

- $\hat{G}$ and $\hat{P}$ are symmetric, i.e.

$$\hat{G}_{ijkl} = \hat{G}_{jikl} = \hat{G}_{ijlk}, \qquad \hat{P}_{ijkl} = \hat{P}_{jikl} = \hat{P}_{ijlk}.$$

- $\hat{P}$ is positive-definite and $\hat{G}$ is non-negative, i.e. for every symmetric second-rank tensor $\mathcal{Z}$ holds:

$$\hat{G}_{ijkl}\mathcal{Z}_{ij}\mathcal{Z}_{kl} \geqslant 0, \qquad \hat{P}_{ijkl}\mathcal{Z}_{ij}\mathcal{Z}_{kl} \geqslant C|\mathcal{Z}|^2, \quad \text{for some constant } C > 0.$$

The computation of tensors $\hat{P}$ and $\hat{G}$ is based on an analytical representation of solutions of the so-called cell equation which arises in homogenization theory. The cell equation is solved numerically by the finite element method. The computation of $\hat{P}$ and $\hat{G}$ is out of scope of this work. For more details see [28].

2.3.2 Mechanical Interface and Boundary Conditions

Figure 2.2 shows the cross section of the biosensor by the plane $x_2 = 0$. The interfaces between the layers are denoted by Γ_1, Γ_2, Γ_3 and Γ_4.

The mechanical conditions at the interfaces between every two contacting solid media include (the homogenized aptamers-fluid layer is considered here as a solid medium):

1. The continuity of the displacement field $\boldsymbol{u}$.

2. The pressure equilibrium, i.e. the continuity of $\sigma \cdot \boldsymbol{n}$, where $\boldsymbol{n}$ is the unit normal vector to the contacting plane. On the interfaces between layers $\boldsymbol{n} = (0,0,1)^T$. Hence this condition is reduced there to the continuity of σ_{i3} for $i = 1,2,3$.

We describe now the mechanical boundary and interface conditions in details.

Interface substrate - guiding layer.

$$\begin{cases} \boldsymbol{u}^p = \boldsymbol{u}^g \\ G_{i3kl}\dfrac{\partial u_l^p}{\partial x_k} + e_{ki3}\dfrac{\partial \varphi^p}{\partial x_k} = \lambda^g \delta_{i3} \mathrm{div}\boldsymbol{u}^g + \mu^g\left(\dfrac{\partial u_i^g}{\partial x_3} + \dfrac{\partial u_3^g}{\partial x_i}\right), \quad i = 1,2,3, \end{cases} \quad \text{on } \Gamma_1. \tag{2.12}$$

Interface guiding layer - schielding layer.

$$\begin{cases} \boldsymbol{u}^g = \boldsymbol{u}^s, \\ \lambda^g \delta_{i3} \mathrm{div} \boldsymbol{u}^g + \mu^g (\frac{\partial u_i^g}{\partial x_3} + \frac{\partial u_3^g}{\partial x_i}) = \lambda^s \delta_{i3} \mathrm{div} \boldsymbol{u}^s + \mu^s (\frac{\partial u_i^s}{\partial x_3} + \frac{\partial u_3^s}{\partial x_i}), \quad i = 1, 2, 3, \end{cases} \quad \text{on } \Gamma_2. \tag{2.13}$$

Interface schielding layer - aptamer.

The homogenized aptamer layer is considered to be solid and therefore the same mechanical conditions take place. The stress in the homogenized aptamer layer is calculated as described in [28]. We have

$$\begin{cases} \boldsymbol{u}^s = \boldsymbol{u}^a, \\ \lambda^s \delta_{i3} \mathrm{div} \boldsymbol{u}^s + \mu^s (\frac{\partial u_i^s}{\partial x_3} + \frac{\partial u_3^s}{\partial x_i}) = \hat{G}^a_{i3kl} \frac{\partial u_l^a}{\partial x_k} + \hat{P}^a_{i3kl} \frac{\partial u_{l\,t}^a}{\partial x_k}, \quad i = 1, 2, 3, \end{cases} \quad \text{on } \Gamma_3. \tag{2.14}$$

Interface aptamer - fluid.

The condition of the continuity of the displacements is replaced here by the no-slip condition that expresses the coupling of oscillations in the aptamer layer and fluid. The requirement of the pressure equilibrium remains. Hence we have

$$\begin{cases} \boldsymbol{u}_t^a = \boldsymbol{v}^f, \\ \hat{G}^a_{i3kl} \frac{\partial u_l^a}{\partial x_k} + \hat{P}^a_{i3kl} \frac{\partial u_{l\,t}^a}{\partial x_k} = -\,p^f \delta_{i3} + \nu \frac{\partial v_i^f}{\partial x_3}, \quad i = 1, 2, 3, \end{cases} \quad \text{on } \Gamma_4. \tag{2.15}$$

Boundary Conditions.

The boundary consists of two components, Γ_0 and Γ_5. The mechanical condition on Γ_0 expresses the absence of force on it, i.e.

$$\sigma_{ij} n_j = 0, \quad i = 1, 2, 3 \qquad \text{on } \Gamma_0, \tag{2.16}$$

where σ is calculated differently in different layers as described in Section 2.3.1 and n_j are components of the unit normal vector $\boldsymbol{n}$ to Γ_5. On the external boundary of Γ_5, we set the no-slip condition for the fluid velocity, i.e.

$$\boldsymbol{v}^f \big|_{\Gamma_5} = 0 \qquad \text{on } \Gamma_5. \tag{2.17}$$

2.3.3 Electrical Boundary Conditions

As mentioned above we assume the electric field in the guiding layer to be negligible. This means that the electric field is involved only in the governing equation for the piezoelectric substrate. The electrical conditions on the boundary surface of the substrate are as follows:

1. On the boundary with the external environment we assume no electric interaction. For this reason the electric flux must be zero there. The same holds for the interface $\Gamma_1 \backslash S$ since electric charges in the guiding layer are neglected. This yields the following Neumann condition:

$$\boldsymbol{D} \cdot \boldsymbol{n} = \left(-\epsilon_{ij} \frac{\partial \varphi}{\partial x_j} + e_{ikl} \frac{\partial u}{\partial x_k} \right) n_i = 0 \qquad \text{on } \partial\Omega_p \setminus S. \tag{2.18}$$

2. The voltage at the electrodes yields Dirichlet boundary conditions. The electrode groups S_1 and S_3 are grounded, and the condition on the input electrodes S_2 expresses the applied voltage inducing an electric field in the substrate, i.e.

$$\varphi(t, \boldsymbol{x})\big|_{S_1} = 0, \tag{2.19}$$

$$\varphi(t, \boldsymbol{x})\big|_{S_2} = V(t)\, \tau^{\varepsilon}(\boldsymbol{x}), \tag{2.20}$$

$$\varphi(t, \boldsymbol{x})\big|_{S_3} = 0, \tag{2.21}$$

where $V(t)$ is a prescribed exciting voltage and $\tau^{\varepsilon}(\boldsymbol{x})$ is a cutoff function satisfying the following condition:

- $\tau^{\varepsilon} \in C_0^{\infty}(S_2)$, $0 \leqslant \tau^{\varepsilon} \leqslant 1$ and $\tau^{\varepsilon} \equiv 1$ outside the ε-neghbourhood of ∂S_2.

Such a function can be constructed for any domain with a Lipschitz boundary. We introduce it here artificially in order to make $\varphi(t, \cdot)\big|_{S_2}$ an element of $C_0^{\infty}(S_2)$. This allows to extend it from S_2 to the whole domain Ω_p as a weak-differentiable function and then to transfer the inhomogenity from the boundary to the right-hand side.

3. The voltage on S_4 as a function of time is the output voltage. It is unknown and to be determined from the solution. However it can not be an arbitrary function; the following restrictions take place. First of all, since the electric field vanishes in the electrodes, the electric potential must remain constant throughout S_4, that is

$$\varphi(t, \boldsymbol{x})\big|_{S_4} = \text{const}(t). \tag{2.22}$$

Futhermore, in contrast to S_1, S_2 and S_3 the voltage on S_4 is not influenced from outside, no charges are brought in or led away. This means that the total electric flux through S_4 must be zero, i.e.

$$\int_{S_4} \boldsymbol{D} \cdot \boldsymbol{n} \, ds = 0. \tag{2.23}$$

2.4 Statement of the Model

2.4.1 Additional Assumptions

In this subsection we introduce some tricks and impose additional assumptions that yield a well-posed model.

Variable transformation.

The main obstacle when deriving a weak formulation of the problem is the no-slip condition at the aptamer-solid interface (the first condition in (2.15)). To overcome this difficulty we apply the method described in [46]. The basic idea is to use the velocity vector instead of the displacement vector in solid. This is achieved by means of the following variable transformation:

$$\boldsymbol{u}(\boldsymbol{x}, t) = \boldsymbol{u}_0(\boldsymbol{x}) + \int_0^t \boldsymbol{v}(\boldsymbol{x}, \tau) d\tau \qquad \text{in } \Omega_{pgsa}. \tag{2.24}$$

Here, $\boldsymbol{v}$ is a new variable describing the velocity of oscillations in the solid layers and $\boldsymbol{u}_0$ is the initial position at the time $t = 0$. This variable trasformation is to perform in all the equations for the solid layers and all the interface and boundary conditions that involve the displacement vector. The no-slip condition in (2.15) converts then into the following natural condition

$$\boldsymbol{v}^a = \boldsymbol{v}^f \qquad \text{on } \Gamma_4.$$

We can consider now the velocity vector $\boldsymbol{v}$ as an unknown funciton defined on the whole domain Ω and continuous on the interfaces between the layers as required by the interface conditions.

In the general case the substitution (2.24) makes the whole system more complicated and requires the specification of the initial displacement $\boldsymbol{u}_0$. We avoid these difficulties by considering only time-periodic solutions.

Time-periodic solutions.

Suppose that ω is the operating frequency of the device. This means that the applied voltage function is of the form

$$V(t) = V_0 \sin \omega t.$$

Then it is natural to assume that the displacements are of the periodic form, i.e.

$$\boldsymbol{u}(\boldsymbol{x}, t) = \boldsymbol{v}^{(1)}(\boldsymbol{x}) \sin \omega t + \boldsymbol{v}^{(2)}(\boldsymbol{x}) \cos \omega t \qquad \text{in } \Omega_{pgsa}. \tag{2.25}$$

The corresponding expression for the velocity is

$$\boldsymbol{v}(\boldsymbol{x}, t) = \omega \boldsymbol{v}^{(1)}(\boldsymbol{x}) \cos \omega t - \omega \boldsymbol{v}^{(2)}(\boldsymbol{x}) \sin \omega t \qquad \text{in } \Omega. \tag{2.26}$$

The same form is assumed for the electric potential and the pressure, i.e.

$$\varphi(\boldsymbol{x}, t) = \varphi^{(1)}(\boldsymbol{x}) \sin \omega t + \varphi^{(2)}(\boldsymbol{x}) \cos \omega t \qquad \text{in } \Omega_p, \tag{2.27}$$

$$p(\boldsymbol{x}, t) = p^{(1)}(\boldsymbol{x}) \sin \omega t + p^{(2)}(\boldsymbol{x}) \cos \omega t \qquad \text{in } \Omega_f. \tag{2.28}$$

Note that (2.25) and (2.26) assume that

$$\boldsymbol{u}_0(\boldsymbol{x}) = -\boldsymbol{v}^{(2)}(\boldsymbol{x}) \qquad \text{in } \Omega_{pgsa}.$$

We can now express $p^{(1)}$ and $p^{(2)}$ through $\boldsymbol{v}^{(1)}$ and $\boldsymbol{v}^{(2)}$ by substituting (2.28) and (2.26) into the second equation in (2.10). This yields

$$\begin{aligned} p^{(1)}(\boldsymbol{x}) &= -\frac{1}{\gamma}\text{div}\boldsymbol{v}^{(1)}, \\ p^{(2)}(\boldsymbol{x}) &= -\frac{1}{\gamma}\text{div}\boldsymbol{v}^{(2)}. \end{aligned} \tag{2.29}$$

Thus, the number of unknown variables is reduced to 4. They are

- $\boldsymbol{v}^{(1)}$ and $\boldsymbol{v}^{(2)}$ in Ω,
- $\varphi^{(1)}$, $\varphi^{(2)}$ in Ω_p.

Dielectric dissipation.

The basic relation between the electric field $\boldsymbol{E}$ and the induced electrical displacement $\boldsymbol{D}$ described by (2.2) is valid for slow processes. As the oscillation frequency grows significantly this relation becomes inaccurate, because the material's polarization does not response to the electric field instantaneously but rather with some time delay. We take it into account as follows.

Denote by $\boldsymbol{D}^P$ the contribution to the electric displacements due to the material's polarization caused by the electric field. This contribution is described in (2.2) by the term $\epsilon_{ij}E_j$. Suppose the electric field oscillates as

$$\boldsymbol{E} = \boldsymbol{E}^{(1)} \sin \omega t.$$

We assume then that $\boldsymbol{D}^P$ oscillates at the same frequency as $\boldsymbol{E}$ but with a small phase lag δ, i.e.

$$\boldsymbol{D}^P = \boldsymbol{D}^{(1)} \sin(\omega t - \delta).$$

Here $\boldsymbol{E}^{(1)}$ and $\boldsymbol{D}^{(1)}$ are amplitudes related by the dielectric permittivity tensor ϵ as in the static case:

$$D_i^{(1)} = \epsilon_{ij} E_j^{(1)}, \quad i = 1, 2, 3.$$

The expression for $\boldsymbol{D}^P$ takes then the form

$$D_i^P = \epsilon_{ij} E_j^{(1)} \cos\delta \, \sin\omega t - \epsilon_{ij} E_j^{(1)} \sin\delta \, \cos\omega t. \tag{2.30}$$

Similarly, the field $\boldsymbol{E} = \boldsymbol{E}^{(2)} \cos\omega t$ causes the electric displacements

$$D_i^P = \epsilon_{ij} E_j^{(2)} \cos\delta \, \cos\omega t + \epsilon_{ij} E_j^{(2)} \sin\delta \, \sin\omega t. \tag{2.31}$$

In our case (2.27) assumes the following form for $\boldsymbol{E}$:

$$\boldsymbol{E} = \boldsymbol{E}^{(1)} \sin\omega t + \boldsymbol{E}^{(2)} \cos\omega t,$$

where $\boldsymbol{E}^{(1)} = -\nabla\varphi^{(1)}$, $\boldsymbol{E}^{(2)} = -\nabla\varphi^{(2)}$. In order to obtain $\boldsymbol{D}^P$ in this case, we combine the contributions described by (2.30) and (2.31). This yields

$$\begin{aligned} D_i^P = {} & \epsilon_{ij} E_j^{(1)} \cos\delta \, \sin\omega t - \epsilon_{ij} E_j^{(1)} \sin\delta \, \cos\omega t + \\ & + \epsilon_{ij} E_j^{(2)} \cos\delta \, \cos\omega t + \epsilon_{ij} E_j^{(2)} \sin\delta \, \sin\omega t = \end{aligned}$$

$$\begin{aligned} &= \left(\epsilon_{ij} E_j^{(1)} \cos\delta + \epsilon_{ij} E_j^{(2)} \sin\delta\right) \sin\omega t + \left(\epsilon_{ij} E_j^{(2)} \cos\delta - \epsilon_{ij} E_j^{(1)} \sin\delta\right) \cos\omega t = \\ &= \left(\epsilon'_{ij} E_j^{(1)} + \epsilon''_{ij} E_j^{(2)}\right) \sin\omega t + \left(\epsilon'_{ij} E_j^{(2)} - \epsilon''_{ij} E_j^{(1)}\right) \cos\omega t. \end{aligned}$$

where $\epsilon' := \epsilon\cos\delta$, $\epsilon'' := \epsilon\sin\delta$. Obviously, both tensors ϵ' and ϵ'' are positive-definite and symmetric. The terms with ϵ' make the main contribution, whereas the terms with ϵ'' are originated from the time delay. The latter are small and describe the energy loss. Rewriting the above relation with $\varphi^{(1)}$ and $\varphi^{(2)}$, we obtain the following expression for the dielectric contribution to the electric displacement:

$$D_i^P = \left(-\epsilon'_{ij}\frac{\partial\varphi^{(1)}}{\partial x_j} - \epsilon''_{ij}\frac{\partial\varphi^{(2)}}{\partial x_j}\right)\sin\omega t + \left(-\epsilon'_{ij}\frac{\partial\varphi^{(2)}}{\partial x_j} + \epsilon''_{ij}\frac{\partial\varphi^{(1)}}{\partial x_j}\right)\cos\omega t. \tag{2.32}$$

In contrast to the expression that we would obtain without taking into account the phase lag, we have here ϵ' instead of ϵ and additional terms with ε'' describing the dissipation of the energy. This correction is taken in consideration when deriving the weak formulation below.

2.4.2 Weak Formulation

We are ready now to derive the basic integral identity. In order to do it we perform the substitutions (2.25)–(2.29) into the governing equations (2.4)–(2.11), equate the coefficients at sine and cosine, multiply the obtained equations by test functions, and integrate them by parts. Summing up all the derived integral identities yields

Contribution of the fluid:

$$-\varrho^f\omega^2\int\limits_{\Omega_f}\boldsymbol{v}^{(1)}\boldsymbol{w}^{(1)}dx + \frac{1}{\gamma}\int\limits_{\Omega_f}\mathrm{div}\boldsymbol{v}^{(1)}\mathrm{div}\boldsymbol{w}^{(1)}dx - \omega\nu^f\int\limits_{\Omega_f}\nabla\boldsymbol{v}^{(2)}:\nabla\boldsymbol{w}^{(1)}dx -$$

$$-\varrho^f\omega^2\int\limits_{\Omega_f}\boldsymbol{v}^{(2)}\boldsymbol{w}^{(2)}dx + \frac{1}{\gamma}\int\limits_{\Omega_f}\mathrm{div}\boldsymbol{v}^{(2)}\mathrm{div}\boldsymbol{w}^{(2)}dx + \omega\nu^f\int\limits_{\Omega_f}\nabla\boldsymbol{v}^{(1)}:\nabla\boldsymbol{w}^{(2)}dx -$$

Contribution of the aptamer layer:

$$-\varrho^a\omega^2\int\limits_{\Omega_a}\boldsymbol{v}^{(1)}\boldsymbol{w}^{(1)}dx + \int\limits_{\Omega_a}\hat{G}\varepsilon(\boldsymbol{v}^{(1)})\varepsilon(\boldsymbol{w}^{(1)})dx - \omega\int\limits_{\Omega_a}\hat{P}\varepsilon(\boldsymbol{v}^{(2)})\varepsilon(\boldsymbol{w}^{(1)})dx -$$

$$-\varrho^a\omega^2\int\limits_{\Omega_a}\boldsymbol{v}^{(2)}\boldsymbol{w}^{(2)}dx + \int\limits_{\Omega_a}\hat{G}\varepsilon(\boldsymbol{v}^{(2)})\varepsilon(\boldsymbol{w}^{(2)})dx + \omega\int\limits_{\Omega_a}\hat{P}\varepsilon(\boldsymbol{v}^{(1)})\varepsilon(\boldsymbol{w}^{(2)})dx -$$

Contribution of the guiding and shielding layers:

$$-\omega^2\int\limits_{\Omega_{gs}}\varrho\boldsymbol{v}^{(1)}\boldsymbol{w}^{(1)}dx + \int\limits_{\Omega_{gs}}\mu\nabla\boldsymbol{v}^{(1)}:\nabla\boldsymbol{w}^{(1)}dx + \int\limits_{\Omega_{gs}}(\lambda+\mu)\mathrm{div}\boldsymbol{v}^{(1)}\mathrm{div}\boldsymbol{w}^{(1)}dx -$$

$$-\omega\int\limits_{\Omega_{gs}^d}\beta(x)\nabla\boldsymbol{v}^{(2)}:\nabla\boldsymbol{w}^{(1)}dx -$$

$$-\omega^2 \int\limits_{\Omega_{gs}} \varrho \boldsymbol{v}^{(2)} \boldsymbol{w}^{(2)} dx + \int\limits_{\Omega_{gs}} \mu \nabla \boldsymbol{v}^{(2)} : \nabla \boldsymbol{w}^{(2)} dx + \int\limits_{\Omega_{gs}} (\lambda + \mu) \mathrm{div} \boldsymbol{v}^{(2)} \mathrm{div} \boldsymbol{w}^{(2)} dx +$$

$$+\omega \int\limits_{\Omega_{gs}^d} \beta(x) \nabla \boldsymbol{v}^{(1)} : \nabla \boldsymbol{w}^{(2)} dx -$$

Mechanical contribution of the substrate:

$$-\varrho^p \omega^2 \int\limits_{\Omega_p} \boldsymbol{v}^{(1)} \boldsymbol{w}^{(1)} dx + \int\limits_{\Omega_p} G \varepsilon(\boldsymbol{v}^{(1)}) \varepsilon(\boldsymbol{w}^{(1)}) dx + \int\limits_{\Omega_p} e_{kij} \frac{\partial \varphi^{(1)}}{\partial x_k} \frac{\partial w_i^{(1)}}{\partial x_j} dx -$$

$$-\omega \int\limits_{\Omega_p^d} \beta(x) \nabla \boldsymbol{v}^{(2)} : \nabla \boldsymbol{w}^{(1)} dx +$$

$$-\varrho^p \omega^2 \int\limits_{\Omega_p} \boldsymbol{v}^{(2)} \boldsymbol{w}^{(2)} dx + \int\limits_{\Omega_p} G \varepsilon(\boldsymbol{v}^{(2)}) \varepsilon(\boldsymbol{w}^{(2)}) dx + \int\limits_{\Omega_p} e_{kij} \frac{\partial \varphi^{(2)}}{\partial x_k} \frac{\partial w_i^{(2)}}{\partial x_j} dx +$$

$$+\omega \int\limits_{\Omega_p^d} \beta(x) \nabla \boldsymbol{v}^{(1)} : \nabla \boldsymbol{w}^{(2)} dx +$$

Electrical contribution of the substrate:

$$+\int\limits_{\Omega_p} \epsilon'_{ij} \frac{\partial \varphi^{(1)}}{\partial x_i} \frac{\partial \psi^{(1)}}{\partial x_j} dx + \int\limits_{\Omega_p} \epsilon''_{ij} \frac{\partial \varphi^{(2)}}{\partial x_i} \frac{\partial \psi^{(1)}}{\partial x_j} dx - \int\limits_{\Omega_p} e_{kij} \frac{\partial v_i^{(1)}}{\partial x_j} \frac{\partial \psi^{(1)}}{\partial x_k} dx +$$

$$+\int\limits_{\Omega_p} \epsilon'_{ij} \frac{\partial \varphi^{(2)}}{\partial x_i} \frac{\partial \psi^{(2)}}{\partial x_j} dx - \int\limits_{\Omega_p} \epsilon''_{ij} \frac{\partial \varphi^{(1)}}{\partial x_i} \frac{\partial \psi^{(2)}}{\partial x_j} dx - \int\limits_{\Omega_p} e_{kij} \frac{\partial v_i^{(2)}}{\partial x_j} \frac{\partial \psi^{(2)}}{\partial x_k} dx =$$

The right hand side:

$$= \int\limits_{\Omega_p} f \psi^{(1)} dx. \tag{2.33}$$

Here $\boldsymbol{w}^{(1)}, \boldsymbol{w}^{(2)}, \psi^{(1)}$ and $\psi^{(2)}$ are test functions. The function f on the right-hand side arises due to the Dirichlet condition (2.20). All the integrals over the interfaces $\Gamma_1, \Gamma_2, \Gamma_3, \Gamma_4$ arising after integrating the mechanical equations by parts express the contribution of the normal components of the stress and, therefore, disappear due to the pressure equilibrium conditions on the interfaces. All the boundary integrals vanish because of the boundary conditions (2.16)–(2.23) and the choice of the test functions. We assume $\boldsymbol{w}^{(1)}, \boldsymbol{w}^{(2)} \in H_{\Gamma_5}$ and $\psi^{(1)}, \psi^{(2)} \in H_S$, where

$$H_{\Gamma_5} := \{\boldsymbol{w} \in H^1(\Omega; \mathbb{R}^3) : \ \boldsymbol{w}\big|_{\Gamma_5} = 0\},$$

$$H_S := \{\psi \in H^1(\Omega_p) : \ \psi\big|_{S_1 \cup S_2 \cup S_3} = 0, \ \psi\big|_{S_4} = \mathrm{const}\}.$$

Both H_{Γ_5} and H_S are complete Hilbert spaces with respect to the inner product defined in $H^1(\Omega;\mathbb{R}^3)$ and $H^1(\Omega_p)$ respectively.

Let us introduce Hilbert spaces $\mathcal{V}$ and $\mathcal{W}$ by

$$\mathcal{V} := (H_{\Gamma_5})^2 \oplus (H_S)^2.$$

$$\mathcal{W} := \left(L^2(\Omega;\mathbb{R}^3)\right)^2 \oplus \left(L^2(\Omega_p)\right)^2.$$

Suppose that $v \in \mathcal{V}$ and $w \in \mathcal{W}$ are of the form

$$v = \left(\boldsymbol{v}^{(1)}, \boldsymbol{v}^{(2)}, \varphi^{(1)}, \varphi^{(2)}\right), \quad w = \left(\boldsymbol{w}^{(1)}, \boldsymbol{w}^{(2)}, \psi^{(1)}, \psi^{(2)}\right).$$

Then norms in $\mathcal{V}$ and $\mathcal{W}$ satisfy

$$\begin{aligned} \|v\|_{\mathcal{V}}^2 &= \|\boldsymbol{v}^{(1)}\|_{H_{\Gamma_5}}^2 + \|\boldsymbol{v}^{(2)}\|_{H_{\Gamma_5}}^2 + \|\varphi^{(1)}\|_{H_S}^2 + \|\varphi^{(2)}\|_{H_S}^2, \\ \|w\|_{\mathcal{W}}^2 &= \|\boldsymbol{w}^{(1)}\|_{L^2(\Omega;\mathbb{R}^3)}^2 + \|\boldsymbol{w}^{(2)}\|_{L^2(\Omega;\mathbb{R}^3)}^2 + \|\psi^{(1)}\|_{L^2(\Omega_p)}^2 + \|\psi^{(2)}\|_{L^2(\Omega_p)}^2. \end{aligned}$$

Remark 2.1. Note that $\mathcal{V}$ is compactly embedded and dense in $\mathcal{W}$ since H_{Γ_5} and H_S are compactly embedded and dense in $L^2(\Omega;\mathbb{R}^3)$ and $L^2(\Omega_p)$ respectively.

Assuming $u, v \in \mathcal{V}$ in the form

$$u = \left(\boldsymbol{v}^{(1)}, \boldsymbol{v}^{(2)}, \varphi^{(1)}, \varphi^{(2)}\right) \in \mathcal{V}, \quad v = \left(\boldsymbol{w}^{(1)}, \boldsymbol{w}^{(2)}, \psi^{(1)}, \psi^{(2)}\right) \in \mathcal{W},$$

we can rewrite the integral identity (2.33) as follows:

$$\tilde{\pi}(u, v) = \tilde{\ell}(v),$$

where $\tilde{\pi}(\cdot,\cdot)$ is a bilinear form on $\mathcal{V} \times \mathcal{V}$ representing the left-hand side of (2.33); $\tilde{\ell}$ is a linear functional on $\mathcal{V}$ standing for the right-hand side. Then the weak formulation of the problem is the following:

Problem 2.1. *Find $u \in \mathcal{V}$ such that*

$$\tilde{\pi}(u, v) = \tilde{\ell}(v) \quad \forall v \in \mathcal{V}.$$

Proposition 2.2. *Let $u = (\boldsymbol{v}^{(1)}, \boldsymbol{v}^{(2)}, \varphi^{(1)}, \varphi^{(2)}) \in \mathcal{V}$ be a solution of Problem 2.2 and the components of u be H^2-functions.*

Then the governing equations (2.4)–(2.11) are fulfilled almost everywhere in the layers. The boundary and interface conditions (2.12)–(2.23) hold almost everywhere on the boundary and the interfaces.

Proof. In order to avoid bulky formulas we provide here only the idea of the proof that is traditional and simple.

First, for every layer we take an arbitrary test function with the support inside the layer and integrate (2.33) by parts. No boundary intergral arises because the support of the test function is inside the layer. After the integration by parts we obtain the governing equation for the layer multiplied by the test function and integrated over the layer. Since the test funciton is arbitrary, the governing equation must hold almost everywhere.

The continuity of the displacements and the conditions (2.16) and (2.22) are fulfilled due to the construction of $\mathcal{V}$. To show that the other interface conditions hold it is enough to take an arbitrary test function with the support in the neighborhood of the interface, integrate (2.33) by parts and use that the governing equations in the layers are fulfilled almost everywhere.

□

Let us decompose the form $\tilde{\pi}(\cdot,\cdot)$ in two parts:

$$\tilde{\pi}(u,v) = \tilde{a}(u,v) - \tilde{b}(u,v),$$

where the form $-\tilde{b}(\cdot,\cdot)$ contains the terms of (2.33) originated from the time derivatives. These are the terms with $\varrho\omega^2$. The form $\tilde{a}(\cdot,\cdot)$ contains all the other terms. Note that the form $\tilde{b}$ is also well-defined on $\mathcal{W}\times\mathcal{W}$ and $\tilde{\ell}$ is well-defined on $\mathcal{W}$. We investigate now the properties of $\tilde{a}(\cdot,\cdot), \tilde{b}(\cdot,\cdot)$ and $\tilde{\ell}$. We will need the following lemma.

Lemma 2.3 (Korn's inequality). *Let Ω be a bounded Lipschitz domain, V a closed subspace of $H^1(\Omega;\mathbb{R}^3)$ and $\Re(\Omega)$ the space of rigid body motions on Ω. If $V\cap\Re(\Omega)=\{0\}$, then there exist a positive constant C depending only on Ω such that for all $\boldsymbol{v}\in V$ holds:*

$$\|\boldsymbol{v}\|_{H^1(\Omega;\mathbb{R}^3)} \leqslant C\|\varepsilon(\boldsymbol{v})\|_{L^2(\Omega;\mathbb{R}^{3\times3})}. \tag{2.34}$$

The proof of Lemma 2.3 can be found for example in [56] (Theorem 2.5).

Proposition 2.4. *The forms $\tilde{a}(\cdot,\cdot)$, $\tilde{b}(\cdot,\cdot)$ and the functional $\tilde{\ell}$ possess the following properties:*

(i) **Boundedness.** *$\tilde{a}(\cdot,\cdot)$ is bounded on $\mathcal{V}\times\mathcal{V}$, $\tilde{b}(\cdot,\cdot)$ is bounded on $\mathcal{W}\times\mathcal{W}$ (and consequently on $\mathcal{V}\times\mathcal{V}$), and $\tilde{\ell}$ is bounded on $\mathcal{W}$ (and on $\mathcal{V}$), i.e. there exist constants c_1, c_2, c_3 such that for all $u, v \in \mathcal{V}$*

$$
\begin{aligned}
\tilde{a}(u,v) &\leqslant c_1\|u\|_{\mathcal{V}}\|v\|_{\mathcal{V}},\\
\tilde{b}(u,v) &\leqslant c_2\|u\|_{\mathcal{W}}\|v\|_{\mathcal{W}} \leqslant c_2\|u\|_{\mathcal{V}}\|v\|_{\mathcal{V}},\\
\tilde{\ell}(v) &\leqslant c_3\|v\|_{\mathcal{W}} \leqslant c_3\|v\|_{\mathcal{V}}.
\end{aligned}
$$

(ii) **Non-negativity.** *$\tilde{a}(\cdot,\cdot)$ and $\tilde{b}(\cdot,\cdot)$ are non-negative. Moreover, there exist a positive constant α such that for all $u = (\boldsymbol{v}^{(1)}, \boldsymbol{v}^{(2)}, \varphi^{(1)}, \varphi^{(2)}) \in \mathcal{V}$ the following estimates hold:*

$$
\tilde{a}((\boldsymbol{v}^{(1)}, \boldsymbol{v}^{(2)}, \varphi^{(1)}, \varphi^{(2)}), (\boldsymbol{v}^{(1)} - \boldsymbol{v}^{(2)}, \boldsymbol{v}^{(2)} + \boldsymbol{v}^{(1)}, \varphi^{(1)} + \varphi^{(2)}, \varphi^{(2)} - \varphi^{(1)})) \geqslant \alpha\|u\|_{\mathcal{V}}, \tag{2.35}
$$
$$
\tilde{b}((\boldsymbol{v}^{(1)}, \boldsymbol{v}^{(2)}, \varphi^{(1)}, \varphi^{(2)}), (\boldsymbol{v}^{(1)} - \boldsymbol{v}^{(2)}, \boldsymbol{v}^{(2)} + \boldsymbol{v}^{(1)}, \varphi^{(1)} + \varphi^{(2)}, \varphi^{(2)} - \varphi^{(1)})) \geqslant 0 \tag{2.36}
$$

Proof. *(i)* The boundedness of $\tilde{a}(\cdot,\cdot)$, $\tilde{b}(\cdot,\cdot)$, and $\tilde{\ell}$ follows from the boundedness of all the terms in (2.33). This can easily be shown by using the Cauchy-Schwarz inequality.

(ii) The estimation (2.36) is obtained trivially from the definition of $\tilde{b}$. To show (2.35) we first show that

$$
\begin{aligned}
&\tilde{a}((\boldsymbol{v}^{(1)}, \boldsymbol{v}^{(2)}, \varphi^{(1)}, \varphi^{(2)}), (\boldsymbol{v}^{(1)}, \boldsymbol{v}^{(2)}, \varphi^{(1)}, \varphi^{(2)})) \geqslant\\
&\quad\geqslant C(\|\nabla \boldsymbol{v}^{(1)}\|^2_{L^2(\Omega_{gs};\mathbb{R}^{3\times3})} + \|\varepsilon(\boldsymbol{v}^{(1)})\|^2_{L^2(\Omega_p;\mathbb{R}^{3\times3})} + \|\varphi^{(1)}\|^2_{H_S}\\
&\quad+\|\nabla \boldsymbol{v}^{(2)}\|^2_{L^2(\Omega_{gs};\mathbb{R}^{3\times3})} + \|\varepsilon(\boldsymbol{v}^{(2)})\|^2_{L^2(\Omega_p;\mathbb{R}^{3\times3})} + \|\varphi^{(2)}\|^2_{H_S})
\end{aligned}
\tag{2.37}
$$

Indeed, the negative terms in $\tilde{a}(u,u)$ have positive counterparts and vanish. The terms originated from the elastic contribution of the piezoelectric are estimated due to the positiveness of G (see (2.6)):

$$
\int_{\Omega_p} G\varepsilon(\boldsymbol{v}^{(1)})\varepsilon(\boldsymbol{v}^{(1)})dx \geqslant C\int_{\Omega_p} \varepsilon(\boldsymbol{v}^{(1)}) : \varepsilon(\boldsymbol{v}^{(1)})dx = C\|\varepsilon(\boldsymbol{v}^{(1)})\|^2_{L^2(\Omega_p;\mathbb{R}^{3\times3})}.
$$

The positiveness of ϵ' and Friedrich's inequality enable us to estimate the electric terms:

$$
\int_{\Omega_p} \epsilon'_{ij}\frac{\partial\varphi^{(1)}}{\partial x_i}\frac{\partial\varphi^{(1)}}{\partial x_j}dx \geqslant C\int_{\Omega_p} |\nabla\varphi^{(1)}|^2dx = C\|\nabla\varphi^{(1)}\|^2_{L^2(\Omega_p)} \geqslant C\|\varphi^{(1)}\|^2_{H_S}.
$$

The terms $\|\nabla \boldsymbol{v}^{(2)}\|^2_{L^2(\Omega_{gs};\mathbb{R}^{3\times 3})}$ and $\|\nabla \boldsymbol{v}^{(1)}\|^2_{L^2(\Omega_{gs};\mathbb{R}^{3\times 3})}$ on the right hand side of (2.37) originate from the contribution of the guiding and shielding layers. They are obtained trivially.

By the same way, using the positiveness of ϵ'' and $\hat{P}$, it can easily be shown that

$$\begin{aligned}\tilde{a}((\boldsymbol{v}^{(1)}, \boldsymbol{v}^{(2)}, \varphi^{(1)}, \varphi^{(2)}),(-\boldsymbol{v}^{(2)}, \boldsymbol{v}^{(1)}, \varphi^{(2)}, -\varphi^{(1)})) \geqslant \\ \geqslant C(\|\nabla \boldsymbol{v}^{(1)}\|^2_{L^2(\Omega_f;\mathbb{R}^{3\times 3})} + \|\varepsilon(\boldsymbol{v}^{(1)})\|^2_{L^2(\Omega_a;\mathbb{R}^{3\times 3})} + \|\varphi^{(1)}\|^2_{H_S} \\ +\|\nabla \boldsymbol{v}^{(2)}\|^2_{L^2(\Omega_f;\mathbb{R}^{3\times 3})} + \|\varepsilon(\boldsymbol{v}^{(2)})\|^2_{L^2(\Omega_a;\mathbb{R}^{3\times 3})} + \|\varphi^{(2)}\|^2_{H_S}).\end{aligned} \tag{2.38}$$

Further, for any domain $\tilde{\Omega} \subset \mathbb{R}^3$ and any $\boldsymbol{u} \in H^1(\tilde{\Omega}; \mathbb{R}^3)$ the following estimate takes place:

$$\|\nabla \boldsymbol{u}\|_{L^2(\tilde{\Omega};\mathbb{R}^3)} \geqslant \|\varepsilon(\boldsymbol{u})\|_{L^2(\tilde{\Omega};\mathbb{R}^3)}. \tag{2.39}$$

Indeed, for any second-rank tensor ξ we have

$$\begin{aligned}\operatorname{sym}(\xi) : \operatorname{sym}(\xi) = \frac{1}{4}(\xi_{ij} + \xi_{ji})(\xi_{ij} + \xi_{ji}) = \frac{1}{2}\xi_{ij}\xi_{ij} + \frac{1}{2}\xi_{ij}\xi_{ji} \leqslant \\ \frac{1}{2}\xi_{ij}\xi_{ij} + \frac{1}{4}\xi_{ij}\xi_{ij} + \frac{1}{4}\xi_{ji}\xi_{ji} = \xi : \xi,\end{aligned}$$

which implies (2.39). We used the Young inequality here.

Adding (2.37) to (2.38) and applying (2.39) to $\boldsymbol{v}^{(1)}$ and $\boldsymbol{v}^{(2)}$ on Ω_{gs}, we obtain

$$\begin{aligned}\tilde{a}((\boldsymbol{v}^{(1)}, \boldsymbol{v}^{(2)}, \varphi^{(1)}, \varphi^{(2)}), (\boldsymbol{v}^{(1)} - \boldsymbol{v}^{(2)}, \boldsymbol{v}^{(2)} + \boldsymbol{v}^{(1)}, \varphi^{(1)} + \varphi^{(2)}, \varphi^{(2)} - \varphi^{(1)})) \geqslant \\ \geqslant C(\|\varepsilon(\boldsymbol{v}^{(1)})\|^2_{L^2(\Omega;\mathbb{R}^{3\times 3})} + \|\varphi^{(1)}\|^2_{H_S} \\ \|\varepsilon(\boldsymbol{v}^{(2)})\|^2_{L^2(\Omega;\mathbb{R}^{3\times 3})} + \|\varphi^{(2)}\|^2_{H_S})\end{aligned}$$

We apply now the Korn inequality (2.34) to $\boldsymbol{v}^{(1)}$ and $\boldsymbol{v}^{(2)}$ as functions from H_{Γ_5} defined on the whole domain Ω and obtain (2.35). We can do this because H_{Γ_5} is a closed subspace of $H^1(\Omega; \mathbb{R}^3)$, and obviously it does not contain any non-zero rigid transformations.

Note that we could not apply the Korn inequality in (2.37) and (2.38) to $\boldsymbol{v}^{(1)}$ and $\boldsymbol{v}^{(2)}$ as functions defined on the subdomains Ω_p and Ω_a, because not all transformations that are rigid locally on Ω_p or Ω_a are necessary excluded from H_{Γ_5}.

□

Let us introduce a linear mapping $Q : \mathcal{V} \to \mathcal{V}$ as follows:

$$Q : (\boldsymbol{w}^{(1)}, \boldsymbol{w}^{(2)}, \psi^{(1)}, \psi^{(2)}) \mapsto (\boldsymbol{w}^{(1)} - \boldsymbol{w}^{(2)}, \boldsymbol{w}^{(2)} + \boldsymbol{w}^{(1)}, \psi^{(1)} + \psi^{(2)}, \psi^{(2)} - \psi^{(1)})$$

Further, for all $u, v \in \mathcal{V}$, let us define

$$\begin{aligned} \pi(u,v) &:= \tilde{\pi}(u,Qv), \\ \ell(v) &:= \tilde{\ell}(Qv), \\ a(u,v) &:= \tilde{a}(u,Qv), \\ b(u,v) &:= \tilde{b}(u,Qv). \end{aligned}$$

Note that by construction

$$\pi(u,v) = a(u,v) - b(u,v).$$

Proposition 2.5. *The forms $a(\cdot,\cdot)$, $b(\cdot,\cdot)$, $\pi(\cdot,\cdot)$ and the functional ℓ possess the following properties:*

(i) **Boundedness**
$a(\cdot,\cdot)$ *is bounded on* $\mathcal{V} \times \mathcal{V}$.
$b(\cdot,\cdot)$ *is bounded on* $\mathcal{W} \times \mathcal{W}$ *and consequently on* $\mathcal{V} \times \mathcal{V}$.
$\pi(\cdot,\cdot)$ *is bounded on* $\mathcal{V} \times \mathcal{V}$.
ℓ *is bounded on* $\mathcal{W}$ *and consequently on* $\mathcal{V}$.

(ii) **Ellipticity and non-negativity**
$a(\cdot,\cdot)$ *is* $\mathcal{V}$*-elliptic.*
$b(\cdot,\cdot)$ *is non-negative on* $\mathcal{W} \times \mathcal{W}$ *(and consequently on* $\mathcal{V} \times \mathcal{V}$*).*

(iii) **Gårding's inequality**
There exist constants $\alpha > 0,\ \beta \in \mathbb{R}$ *such that*

$$\pi(u,\, u) \geqslant \alpha \|u\|_{\mathcal{V}}^2 - \beta \|u\|_{\mathcal{W}}^2 \qquad \forall\, u \in \mathcal{V}.$$

Proof. The statements (i) and (ii) follow from Proposition 2.4 and the definitions of a, b, l, and π. The property (ii) is derived directly. In order to show (i), it suffices to prove that

$$\|Qu\|_{\mathcal{W}} \leqslant C\|u\|_{\mathcal{W}} \quad \forall u \in \mathcal{W} \qquad \text{and} \qquad \|Qu\|_{\mathcal{V}} \leqslant C\|u\|_{\mathcal{V}} \quad \forall u \in \mathcal{V}$$

with some appropriate positive constant C.
Let $u \in \mathcal{V}$ be of the form $u = (\boldsymbol{v}^{(1)}, \boldsymbol{v}^{(2)}, \varphi^{(1)}, \varphi^{(2)})$. Then

$$\|Qu\|_{\mathcal{V}}^2 = \|\boldsymbol{v}^{(1)}\|_{H_{\Gamma_5}}^2 + \|\boldsymbol{v}^{(2)}\|_{H_{\Gamma_5}}^2 + \|\varphi^{(1)}\|_{H_S}^2 + \|\varphi^{(1)}\|_{H_S}^2 =$$

$$= \|\boldsymbol{v}^{(1)} - \boldsymbol{v}^{(2)}\|^2_{H_{\Gamma_5}} + \|\boldsymbol{v}^{(2)} + \boldsymbol{v}^{(1)}\|^2_{H_{\Gamma_5}} + \|\varphi^{(1)} + \varphi^{(2)}\|^2_{H_S} + \|\varphi^{(2)} - \varphi^{(1)}\|^2_{H_S} \leqslant$$
$$\leqslant 2(\|\boldsymbol{v}^{(1)}\|_{H_{\Gamma_5}} + \|\boldsymbol{v}^{(2)}\|_{H_{\Gamma_5}})^2 + 2(\|\varphi^{(1)}\|_{H_S} + \|\varphi^{(2)}\|_{H_S})^2 \leqslant$$
$$\leqslant 4\|\boldsymbol{v}^{(1)}\|^2_{H_{\Gamma_5}} + 4\|\boldsymbol{v}^{(2)}\|^2_{H_{\Gamma_5}} + 4\|\varphi^{(1)}\|^2_{H_S} + 4\|\varphi^{(2)}\|^2_{H_S} = 4\|u\|^2_{\mathcal{V}}.$$

Hence,

$$\|Qu\|_{\mathcal{V}} \leqslant 2\|u\|_{\mathcal{V}}.$$

It can be shown in the same way that $\|Qu\|_{\mathcal{W}} \leqslant 2\|u\|_{\mathcal{W}}$. Therefore the boundedness of $\tilde{a}$, $\tilde{b}$, $\tilde{\pi}$ and $\tilde{\ell}$ implies the boundedness of a, b, π and ℓ.

The statement *(iii)* follows straightforwardly from the $\mathcal{V}$-ellipticity of $a(\cdot,\cdot)$ and boundedness of $b(\cdot,\cdot)$ on $\mathcal{W} \times \mathcal{W}$.

□

Let us formulate the problem in terms of the new forms.

Problem 2.2. *Find $u \in \mathcal{V}$ such that*

$$\pi(u, v) = \ell(v) \quad \forall v \in \mathcal{V}.$$

Proposition 2.6. *Problem 2.1 and Problem 2.2 are equivalent, i.e. u is a solution of Problem 2.1 iff u is a solution of Problem 2.2.*

Proof. It easy to see that the mapping Q admits the representation:

$$(Qv)^T = \begin{pmatrix} 1 & 1 & 0 & 0 \\ 1 & 1 & 0 & 0 \\ 0 & 0 & 1 & 1 \\ 0 & 0 & -1 & 1 \end{pmatrix} v^T$$

Since the matrix in this representation is non-singular, we can construct a reverse mapping defined on the whole $\mathcal{V}$. This implies that Q is a bijective transformation on $\mathcal{V}$. Therefore, if an identity holds for all $\{Qv \mid v \in \mathcal{V}\}$, it must also hold for all $v \in \mathcal{V}$ and vice verse.

□

2.5 Well-Posedness of the Model

In this section we discuss the well-posedness of the problem stated above in Hadamard's sense. We establish that Problem 2.2 has a unique solution that depends continuously on the boundary conditions represented by the functional ℓ. The proof is based on a generalization of the Lax-Milgram theorem.

Definition. Let $\mathcal{V}$ be a normed space, $a(\cdot,\cdot)$ a continuous bilinear form on $\mathcal{V}\times\mathcal{V}$. We say that the operator $A \in L(\mathcal{V},\mathcal{V}')$ is *associated to* $a(\cdot,\cdot)$ if

$$a(u,\, v) = \langle Au,\, v\rangle_{\mathcal{V}',\mathcal{V}} \qquad \text{for all } u,\, v \in \mathcal{V}.$$

Such an operator always exists and acts as follows. To every fixed $u \in \mathcal{V}$ it assigns $a(u,\cdot)$ considered as a functional on $\mathcal{V}$. The continuity of A follows directly from the continuity of $a(\cdot,\cdot)$.

Denoting by A the operator associated to $\pi(\cdot,\cdot)$ we can reformulate Problem 2.2 as an equation

$$Au = \ell. \tag{2.40}$$

The well-posedness of the model is then equivalent to the statement that A is inversible and the inverse operator A^{-1} is defined and bounded on the whole space $\mathcal{V}'$.

Theorem 2.7. *Let $\mathcal{V}$ be a Hilbert space and $A \in L(\mathcal{V},\mathcal{V}')$ be the operator associated to a continuous bilinear form $\pi(\cdot,\cdot)$ on $\mathcal{V}\times\mathcal{V}$. Further, let $\pi(\cdot,\cdot)$ satisfy*

$$\inf_{u\in S_{\mathcal{V}}} \sup_{v\in S_{\mathcal{V}}} \pi(u,v) \;=:\; \epsilon > 0, \tag{2.41}$$

$$\inf_{v\in S_{\mathcal{V}}} \sup_{u\in S_{\mathcal{V}}} \pi(u,v) \;=:\; \epsilon' > 0, \tag{2.42}$$

where $S_{\mathcal{V}} := \{v \in \mathcal{V} :\ \|v\|_{\mathcal{V}} = 1\}$ is the unit sphere in $\mathcal{V}$.

Then

- $\epsilon = \epsilon'$,
- $A^{-1} \in L(\mathcal{V}',\mathcal{V})$ *exists and* $\|A^{-1}\| = \dfrac{1}{\epsilon}$.

The theorem is based on the result of Nečas (see [53]). The proof of this theorem in English can be found for example in [74] (Theorem 6.5.9).

Thus if the conditions (2.41) and (2.42) are fullfiled, Problem 2.2 has a unique solution u for every $\ell \in \mathcal{V}'$, and it satisfies

$$\|u\|_{\mathcal{V}} \leqslant \frac{1}{\epsilon}\|\ell\|_{\mathcal{V}'}.$$

The following lemma enables us to get rid of the condition (2.42).

Lemma 2.8. *Let $\mathcal{V}$ and $\mathcal{W}$ be Hilbert spaces with compact dense embedding $\mathcal{V} \subset \mathcal{W}$. Let π be a continuous bilinear form on $\mathcal{V} \times \mathcal{V}$ satisfying*

$$\pi(u,\, u) \geqslant \alpha\|u\|_{\mathcal{V}}^2 - \beta\|u\|_{\mathcal{W}}^2 \qquad \forall\, u \in \mathcal{V} \tag{2.43}$$

with some appropriate constants $\alpha > 0$ and $\beta \in \mathbb{R}$.
Then the conditions (2.41) and (2.42) are equivalent.

See [74] (Lemma 6.5.17) for the proof.

In our case, all the conditions of the lemma are satisfied. Indeed, $\mathcal{V}$ is compactly embedded and dense in $\mathcal{W}$ by Remark 2.1. The inequality (2.43) was established in Proposition 2.5. Therefore, in order to prove the well-posedness of Problem 2.2, it suffices to show (2.41).

Remark 2.9. Let us consider the problem adjont to Problem 2.2. It reads:
Find $u \in \mathcal{V}$ such that

$$\pi^*(u,\, v) = \ell(v) \qquad \forall\, v \in \mathcal{V},$$

where $\pi^*(u,\, v) := \pi(v,\, u)$ is the form adjont to $\pi(\cdot,\cdot)$. Obviously Theorem 2.7 and Lemma 2.8 are applicable to π^* to the same extent as to π. The only diffrence is that the conditions (2.41) and (2.42) get swapped. However, since these two conditions are equivalent by Lemma 2.8, the condition (2.41) guarantees the well-posedness of the adjont problem as well.

Problem 2.3. *Find $w \in \mathcal{V}$ such that*

$$\pi(w, v) = 0 \quad \forall\, v \in \mathcal{V}.$$

Theorem 2.10. *Let $w = 0$ be a unique solution of Problem 2.3. Then the condition (2.41)*

$$\inf_{u \in S_{\mathcal{V}}} \sup_{v \in S_{\mathcal{V}}} \pi(u, v) > 0$$

is fulfilled.

Proof. Suppose the claim is false. Then there exist sequences $\mu^m \to 0$ and $u^m \in S_\mathcal{V}$ such that

$$\sup_{v \in S_\mathcal{V}} \pi(u^m, v) < \mu^m.$$

That is

$$\sup_{v \in S_\mathcal{V}} [a(u^m, v) - b(u^m, v)] < \mu^m. \tag{2.44}$$

Suppose that there exist $M \in \mathbb{N}$ such that for all $m > M$

$$b(u^m, u^m) = 0.$$

Substituting this into (2.44) and using the coercivity of a, we obtain

$$\alpha = \alpha \|u^m\|_\mathcal{V}^2 \leqslant a(u^m, u^m) < \mu^m,$$

where α is the ellipticity constant of $a(\cdot, \cdot)$. This inequality implies that $\alpha = 0$, which is a contradiction since α is positive by definition. We can therefore extract a subsequence of u^m, that we denote by the same index m, such that

$$b(u^m, u^m) > 0 \quad \forall m \in \mathbb{N}.$$

We can then introduce

$$w^m := \frac{u^m}{\sqrt{b(u^m, u^m)}} \neq 0.$$

Note that by construction

$$b(w^m, w^m) = 1 \quad \text{and} \quad \|w^m\|_\mathcal{V} = \frac{1}{\sqrt{b(u^m, u^m)}}.$$

Multiplying (2.44) by $\dfrac{1}{\sqrt{b(u^m, u^m)}}$ yields

$$\sup_{v \in S_\mathcal{V}} [a(w^m, v) - b(w^m, v)] < \mu^m \|w^m\|_\mathcal{V}. \tag{2.45}$$

In particular, for $v = \dfrac{w^m}{\|w^m\|_\mathcal{V}}$ we have

$$a(w^m, w^m) - 1 < \mu^m \|w^m\|_\mathcal{V}^2.$$

The $\mathcal{V}$-ellipticity of $a(\cdot,\cdot)$ implies

$$(\alpha - \mu^m)\|w^m\|_{\mathcal{V}}^2 < 1.$$

Since $\mu_m \to 0$, this estimate means that w_m is bounded in $\mathcal{V}$. Therefore it contains a subsequence that converges weakly in $\mathcal{V}$ to some limit w^0. We denote the subsequence by the same index m. Recall that $\mathcal{W}$ is compactly embedded in $\mathcal{V}$ and therefore

$$w^m \to w^0 \quad \text{in } \mathcal{W}.$$

Since b is bounded on $\mathcal{W} \times \mathcal{W}$, this implies that

$$b(w^m, w^m) \to b(w^0, w^0)$$

and hence

$$b(w^0, w^0) = 1.$$

This means that $w^0 \neq 0$. We would like to show now that w_0 solves Problem 2.3.

Let $\{v^n\} \subset S_{\mathcal{V}}$ be a sequence such that $\pi(w^0, v^n) \to \sup\limits_{v \in S_{\mathcal{V}}} \pi(w^0, v)$ as $n \to \infty$. Since $\mathcal{V}$ is reflexiv and $S_{\mathcal{V}}$ is bounded, we can extract a subsequence of $\{v^{n_k}\}$ that converges weakly to some $v^0 \in \mathcal{V}$. Note that v^0 does not have to belong to $S_{\mathcal{V}}$ because $S_{\mathcal{V}}$ is not closed in the weak topology. Further, $\pi(w^0, v)$ at fixed w^0 can be considered as a bounded linear functional on $\mathcal{V}$ and therefore $\pi(w^0, v^{n_k}) \to \pi(w^0, v^0)$. Hence $\pi(w^0, v^0) = \sup\limits_{v \in S_{\mathcal{V}}} \pi(w^0, v)$. Then (2.45) implies

$$\pi(w^m, v^0) = \sup_{v \in S_{\mathcal{V}}} \pi(w^m, v) < \mu^m \|w^m\|_{\mathcal{V}}.$$

Taking the limit as $m \to \infty$ yields

$$\pi(w^0, v^0) \leqslant 0.$$

By construction of v^0 this implies that

$$\sup_{v \in S_{\mathcal{V}}} \pi(w^0, v) \leqslant 0.$$

Hence,

$$\pi(w^0, v) = 0 \quad \forall\, v \in S_{\mathcal{V}}.$$

Therefore w^0 is a solution of Problem 2.3. By construction $w^0 \neq 0$. This contradicts the condition of the theorem.

□

Thus, in order to ensure the well-posedness of the model, it is left to show that the homogeneous problem has only the trivial solution. We will need the following well-known results.

Definition. Let $\boldsymbol{e}_m$ be the m-th basis vector, and suppose that $h \in \mathbb{R} \setminus \{0\}$, $U, V \subset \mathbb{R}^n$ are such that

$$V + h\boldsymbol{e}_m \subset U.$$

Further, let w be a scalar or vector function on U. Then the function

$$D_m^h w(\boldsymbol{x}) := \frac{w(\boldsymbol{x} + h\boldsymbol{e}_m) - w(\boldsymbol{x})}{h}, \quad x \in V.$$

is called the *m-th difference quotient of size h.*

Theorem 2.11. *Let $\boldsymbol{e}_m$ be the m-th basis vector, and suppose that $\delta > 0$, open domains $U, V \subset \mathbb{R}^n$ are such that*

$$V \pm \delta\boldsymbol{e}_m \subset U.$$

(i) Let $w \in L^2(U)$ and $\dfrac{\partial w}{\partial x_m} \in L^2(U)$. Then there exists a constant C such that

$$\left\| D_m^h w \right\|_{L^2(V)} \leqslant C \left\| \frac{\partial w}{\partial x_m} \right\|_{L^2(U)} \tag{2.46}$$

for all $0 < |h| < \dfrac{\delta}{2}$. The constant C does not depend on w.

(ii) Assume $w \in L^2(U)$, and there exists a constant C such that

$$\left\| D_m^h w \right\|_{L^2(V)} \leqslant C$$

for all $0 < |h| < \dfrac{\delta}{2}$. Then

$$\frac{\partial w}{\partial x_m} \in L^2(V) \quad \text{and} \quad \left\| \frac{\partial w}{\partial x_m} \right\|_{L^2(V)} \leqslant C.$$

The proof of these statements can be found for example in [18].

We are ready now to prove the main theorem.

Theorem 2.12. *Suppose* $u = (\boldsymbol{v}^{(1)}, \boldsymbol{v}^{(2)}, \varphi^{(1)}, \varphi^{(2)}) \in \mathcal{V}$ *is a solution of Problem 2.3, i.e.*

$$\pi(u, v) = 0 \qquad \forall v \in \mathcal{V}. \tag{2.47}$$

Then $u = 0$.

Proof. By definition of π (see also Proposition 2.6), (2.47) is equivalent to

$$\tilde{\pi}(u, v) = 0 \qquad \forall\, v \in \mathcal{V}. \tag{2.48}$$

Our goal is to prove that (2.48) implies that $(\boldsymbol{v}^{(1)}, \boldsymbol{v}^{(2)})$ and $(\varphi^{(1)}, \varphi^{(2)})$ vanish in Ω and Ω_p respectively. We do it step by step for subdomains of Ω.

Recall that $\tilde{\pi}$ is the bilinear form representing the left-hand side of the integral identity (2.33). Taking v in the form $(-\boldsymbol{v}^{(2)}, \boldsymbol{v}^{(1)}, \varphi^{(2)}, -\varphi^{(1)})$ and substituting it into (2.48), we obtain

$$\begin{aligned} 0 = \tilde{\pi}(u, v) = \omega\nu^f \int_{\Omega_f} \nabla\boldsymbol{v}^{(2)} : \nabla\boldsymbol{v}^{(2)} dx + \omega\nu^f \int_{\Omega_f} \nabla\boldsymbol{v}^{(1)} : \nabla\boldsymbol{v}^{(1)} dx + \\ + \omega \int_{\Omega_a} \hat{P}\varepsilon(\boldsymbol{v}^{(2)})\varepsilon(\boldsymbol{v}^{(2)}) dx + \omega \int_{\Omega_a} \hat{P}\varepsilon(\boldsymbol{v}^{(1)})\varepsilon(\boldsymbol{v}^{(1)}) dx + \\ + \omega \int_{\Omega_{pgs}^d} \beta(x)\nabla\boldsymbol{v}^{(2)} : \nabla\boldsymbol{v}^{(2)} dx + \omega \int_{\Omega_{pgs}^d} \beta(x)\nabla\boldsymbol{v}^{(1)} : \nabla\boldsymbol{v}^{(1)} dx + \\ + \int_{\Omega_p} \epsilon_{ij}'' \frac{\partial\varphi^{(2)}}{\partial x_i}\frac{\partial\varphi^{(2)}}{\partial x_j} dx + \int_{\Omega_p} \epsilon_{ij}'' \frac{\partial\varphi^{(1)}}{\partial x_i}\frac{\partial\varphi^{(1)}}{\partial x_j} dx. \end{aligned} \tag{2.49}$$

Since all the terms on the right-hand side are non-negative and the sum of them is zero, each of them must be zero. Hence for the first two terms we have

$$\omega\nu^f \int_{\Omega_f} \nabla\boldsymbol{v}^{(2)} : \nabla\boldsymbol{v}^{(2)} dx = \omega\nu^f \int_{\Omega_f} \nabla\boldsymbol{v}^{(1)} : \nabla\boldsymbol{v}^{(1)} dx = 0$$

This implies $\nabla\boldsymbol{v}^{(1)} \equiv \nabla\boldsymbol{v}^{(2)} \equiv 0$ in Ω_f. By definition of $\mathcal{V}$ $\boldsymbol{v}^{(1)}\big|_{\Gamma_5}, \boldsymbol{v}^{(2)}\big|_{\Gamma_5} = 0$. Hence $\boldsymbol{v}^{(1)} \equiv \boldsymbol{v}^{(2)} \equiv 0$ in Ω_f.

Since $\boldsymbol{v}^{(1)}$ and $\boldsymbol{v}^{(2)}$ are H^1-functions, their traces on $\partial\Omega_f$ must be zero as well. In particular $\boldsymbol{v}^{(1)}$ and $\boldsymbol{v}^{(2)}$ are zero at the interface between the fluid (Ω_f) and the aptamer (Ω_a) layers. This enables us to use Korn's inequality when treating the second pair of terms from (2.49). We have

$$0 = \omega \int\limits_{\Omega_a} \hat{P}\varepsilon(\boldsymbol{v}^{(i)})\varepsilon(\boldsymbol{v}^{(i)})dx \geqslant C\omega \int\limits_{\Omega_a} \varepsilon(\boldsymbol{v}^{(i)}) : \varepsilon(\boldsymbol{v}^{(i)})dx =$$

$$= C\omega \|\varepsilon(\boldsymbol{v}^{(i)})\|_{L^2(\Omega_a;\mathbb{R}^{3\times 3})} \geqslant C_1 \|\boldsymbol{v}^{(i)}\|_{H^1(\Omega_a;\mathbb{R}^3)} \geqslant 0, \quad i = 1, 2.$$

Therefore $\boldsymbol{v}^{(1)} \equiv \boldsymbol{v}^{(2)} \equiv 0$ in Ω_a.

Let us now consider the third pair of terms from the right-hand side of (2.49). They satisfy

$$\omega \int\limits_{\Omega^d_{pgs}} \beta(x)\nabla\boldsymbol{v}^{(2)} : \nabla\boldsymbol{v}^{(2)}dx = \omega \int\limits_{\Omega^d_{pgs}} \beta(x)\nabla\boldsymbol{v}^{(1)} : \nabla\boldsymbol{v}^{(1)}dx = 0.$$

Since $\beta(x)$ is positive in Ω^d_{pgs} by definition, these equalities imply $\nabla\boldsymbol{v}^{(1)} \equiv \nabla\boldsymbol{v}^{(2)} \equiv 0$ in Ω^d_{pgs}. At the same time $\boldsymbol{v}^{(1)}$ and $\boldsymbol{v}^{(2)}$ are H^1-functions and they vanish in the adjacent domain Ω_a. Consequently $\boldsymbol{v}^{(1)} = \boldsymbol{v}^{(2)} = 0$ in Ω^d_{pgs}.

Finally, for the last two terms, we have

$$0 = \int\limits_{\Omega_p} \epsilon''_{ij} \frac{\partial\varphi^{(i)}}{\partial x_i}\frac{\partial\varphi^{(i)}}{\partial x_j} dx \geqslant C \int\limits_{\Omega_p} \left|\nabla\varphi^{(i)}\right|^2 dx = C \left\|\nabla\varphi^{(i)}\right\|_{L^2(\Omega_p;\mathbb{R}^3)} \geqslant 0, \quad i = 1, 2.$$

This implies $\nabla\varphi^{(1)} \equiv \nabla\varphi^{(2)} \equiv 0$ in Ω_p. We use now that $\varphi^{(1)}$, $\varphi^{(2)} \in H_S$ and therefore, by definition of H_S, $\varphi^{(1)}\big|_S = \varphi^{(2)}\big|_S = 0$. Hence $\varphi^{(1)} \equiv \varphi^{(2)} \equiv 0$.

Thus, we have proven that

$$\begin{aligned} \boldsymbol{v}^{(1)} = \boldsymbol{v}^{(2)} = 0 \quad & \text{in } \Omega_f \cup \Omega_a \cup \Omega^d_{pgs}, \\ \varphi^{(1)} = \varphi^{(2)} = 0 \quad & \text{in } \Omega_p. \end{aligned} \tag{2.50}$$

It is left to show that $\boldsymbol{v}^{(1)}$ and $\boldsymbol{v}^{(2)}$ vanish on Ω_s, Ω_g and Ω_p (see Figure 2.3).

Due to (2.50) the integral identity (2.33) decouples into four independent equations for $\boldsymbol{v}^{(1)}$ and for $\boldsymbol{v}^{(2)}$. These two pairs of equations are completely identical and not connected

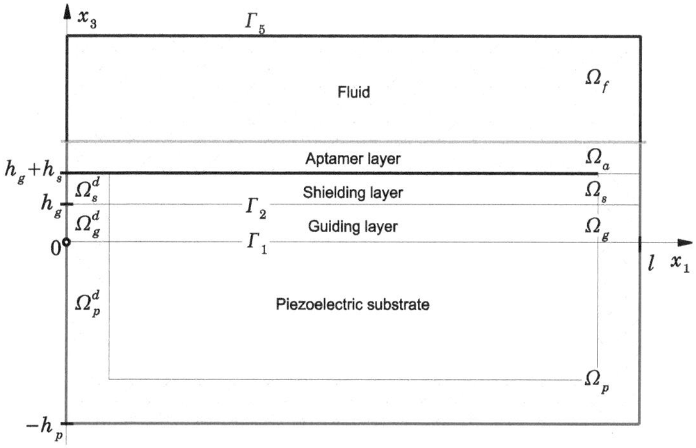

Figure 2.3: Cross section of the domain. The subdomain where $\boldsymbol{v} = 0$ is hatched.

by test functions. The equations are

$$\begin{aligned}
&-\varrho^s\omega^2 \int\limits_{\Omega_s} \boldsymbol{v}\cdot\boldsymbol{w}\,dx + \mu^s \int\limits_{\Omega_s} \nabla\boldsymbol{v} : \nabla\boldsymbol{w}\,dx \ + (\lambda^s+\mu^s)\int\limits_{\Omega_s} \operatorname{div}\boldsymbol{v}\operatorname{div}\boldsymbol{w}\,dx - \\
&-\varrho^g\omega^2 \int\limits_{\Omega_g} \boldsymbol{v}\cdot\boldsymbol{w}\,dx + \mu^g \int\limits_{\Omega_g} \nabla\boldsymbol{v} : \nabla\boldsymbol{w}\,dx \ + (\lambda^g+\mu^g)\int\limits_{\Omega_g} \operatorname{div}\boldsymbol{v}\operatorname{div}\boldsymbol{w}\,dx - \\
&-\varrho^p\omega^2 \int\limits_{\Omega_p} \boldsymbol{v}\cdot\boldsymbol{w}\,dx + \int\limits_{\Omega_p} G\varepsilon(\boldsymbol{v})\varepsilon(\boldsymbol{w})\,dx \ = 0 \qquad \forall\,\boldsymbol{w}\in H_{\Gamma_5},
\end{aligned} \tag{2.51}$$

$$\int\limits_{\Omega_p} e_{kij}\frac{\partial v_i}{\partial x_j}\frac{\partial \psi}{\partial x_k}dx = 0 \qquad \forall\,\psi\in H_S. \tag{2.52}$$

The function $\boldsymbol{v}$ here is either $\boldsymbol{v}^{(1)}$ or $\boldsymbol{v}^{(2)}$. Since both $\boldsymbol{v}^{(1)}$ and $\boldsymbol{v}^{(2)}$ are determined by the same equations, we will not differ them and write from now on just $\boldsymbol{v}$ meaning $\boldsymbol{v}^{(1)}$ or $\boldsymbol{v}^{(2)}$.

We show now that (2.51) and the fact that $\boldsymbol{v} = 0$ in the damping domain $\Omega^d_{pgs}(:=\operatorname{int}\overline{\Omega^d_p\cup\Omega^d_g\cup\Omega^d_s})$ imply that $\boldsymbol{v}=0$ in $\Omega_{pgs}(:=\operatorname{int}\overline{\Omega_p\cup\Omega_g\cup\Omega_s})$.

The first step is to prove that for all $n\in\mathbb{N}_0$ the following statements hold:

(i) $\dfrac{\partial^n \boldsymbol{v}}{\partial x_1^n} \in H^1(\Omega_{pgs}; \mathbb{R}^3)$.

(ii) There is a constant C independent on n such that

$$\left\| \frac{\partial^n \boldsymbol{v}}{\partial x_1^n} \right\|_{L^2(\Omega_{pgs};\mathbb{R}^3)} \leqslant C^n \left\| \boldsymbol{v} \right\|_{L^2(\Omega_{pgs};\mathbb{R}^3)} . \tag{2.53}$$

We prove (i) by induction and while doing it we derive an estimate that proves (ii).

Basis. For $n = 0$ the statement (i) holds because $\boldsymbol{v} \in H_{\Gamma_5} \subset H^1(\Omega; \mathbb{R}^3)$.

Inductive step. Assume now that (i) is fullfiled for some $n \in \mathbb{N}_0$ and let us prove that this implies

$$\frac{\partial^{n+1} \boldsymbol{v}}{\partial x_1^{n+1}} \in H^1(\Omega_{pgs}; \mathbb{R}^3).$$

Let us first show that $\dfrac{\partial^n \boldsymbol{v}}{\partial x_1^n}$ in place of $\boldsymbol{v}$ satisfies the integral identity (2.51). Note that we can put it there because by the induction hypothesis $\dfrac{\partial^n \boldsymbol{v}}{\partial x_1^n} \in H^1(\Omega_{pgs}; \mathbb{R}^3)$. Taking an arbitrary $\boldsymbol{w} \in C^\infty(\overline{\Omega_{pgs}}; \mathbb{R}^3)$, substituting $(-1)^n \dfrac{\partial^n \boldsymbol{w}}{\partial x_1^n}$ into (2.51) as a test function and integrating n times by parts yields

$$\begin{aligned}
&\mu^s \int\limits_{\Omega_s} \nabla \frac{\partial^n \boldsymbol{v}}{\partial x_1^n} : \nabla \boldsymbol{w}\, dx + (\lambda^s + \mu^s) \int\limits_{\Omega_s} \operatorname{div} \frac{\partial^n \boldsymbol{v}}{\partial x_1^n} \operatorname{div} \boldsymbol{w} dx + \\
&+\mu^g \int\limits_{\Omega_g} \nabla \frac{\partial^n \boldsymbol{v}}{\partial x_1^n} : \nabla \boldsymbol{w}\, dx + (\lambda^g + \mu^g) \int\limits_{\Omega_g} \operatorname{div} \frac{\partial^n \boldsymbol{v}}{\partial x_1^n} \operatorname{div} \boldsymbol{w} dx + \\
&+ \int\limits_{\Omega_p} G\varepsilon \left(\frac{\partial^n \boldsymbol{v}}{\partial x_1^n} \right) \varepsilon(\boldsymbol{w}) dx = \\
&\qquad = \varrho^s \omega^2 \int\limits_{\Omega_s} \frac{\partial^n \boldsymbol{v}}{\partial x_1^n} \cdot \boldsymbol{w}\, dx + \varrho^g \omega^2 \int\limits_{\Omega_g} \frac{\partial^n \boldsymbol{v}}{\partial x_1^n} \cdot \boldsymbol{w}\, dx + \varrho^p \omega^2 \int\limits_{\Omega_p} \frac{\partial^n \boldsymbol{v}}{\partial x_1^n} \cdot \boldsymbol{w}\, dx.
\end{aligned} \tag{2.54}$$

Note that no surface integrals arise during integration by parts. The integrals over the side surfaces vanish because $\boldsymbol{v} = 0$ in Ω^d_{pgs}; the integrals over the top and bottom surfaces are zero because the differentiation direction is orthogonal to the normal vectors. Since (2.54) holds for all $\boldsymbol{w} \in C^\infty(\overline{\Omega_{pgs}}; \mathbb{R}^3)$, it must also hold for all $\boldsymbol{w} \in H^1(\Omega_{pgs}; \mathbb{R}^3)$ because $C^\infty(\overline{\Omega_{pgs}}; \mathbb{R}^3)$ is dense in $H^1(\Omega_{pgs}; \mathbb{R}^3)$.

Denote by h_p, h_g, and h_s the thickness of the piezoelectric, guiding, and shielding layers, respectively. Put

$$h_{gs} := h_g + h_s.$$

Then by the choice of the origin of coordinates

$$\Omega_{pgs} = (0, l) \times (0, w) \times (-h_p, h_{gs}),$$

where l and w are the length and the width of the biosensor respectively. Further, denote by δ the thickness of the damping domain along x_1-axis. Let

$$\boldsymbol{w}(x) := -D_1^{-h}\left(D_1^h \frac{\partial^n \boldsymbol{v}}{\partial x_1^n}\right), \quad x \in \left(\frac{\delta}{2}, l - \frac{\delta}{2}\right) \times (0, w) \times (-h_p, h_{gs}),$$

where $h < \dfrac{\delta}{2}$ and D_1^h is the difference quotient of size h with respect to x_1. Since $\boldsymbol{v} = 0$ for $x_1 \in (0, \delta] \cup [l - \delta, l)$, $\boldsymbol{w}$ can be continuously extended by 0 on Ω_{pgs} and further on the whole domain Ω. By the induction hypothesis $\dfrac{\partial^n \boldsymbol{v}}{\partial x_1^n} \in H^1(\Omega_{pgs}; \mathbb{R}^3)$ and hence $\boldsymbol{w}$ belongs to $H^1(\Omega_{pgs}; \mathbb{R}^3)$ too. Therefore, we can use $\boldsymbol{w}$ as a test function in (2.54). This yields

$$\begin{aligned}
&-\mu^s \int_{\Omega_s} \nabla \frac{\partial^n \boldsymbol{v}}{\partial x_1^n} : \nabla D_1^{-h}\left(D_1^h \frac{\partial^n \boldsymbol{v}}{\partial x_1^n}\right) dx - (\lambda^s + \mu^s) \int_{\Omega_s} \operatorname{div} \frac{\partial^n \boldsymbol{v}}{\partial x_1^n} \operatorname{div} D_1^{-h}\left(D_1^h \frac{\partial^n \boldsymbol{v}}{\partial x_1^n}\right) dx - \\
&-\mu^g \int_{\Omega_g} \nabla \frac{\partial^n \boldsymbol{v}}{\partial x_1^n} : \nabla D_1^{-h}\left(D_1^h \frac{\partial^n \boldsymbol{v}}{\partial x_1^n}\right) dx - (\lambda^g + \mu^g) \int_{\Omega_g} \operatorname{div} \frac{\partial^n \boldsymbol{v}}{\partial x_1^n} \operatorname{div} D_1^{-h}\left(D_1^h \frac{\partial^n \boldsymbol{v}}{\partial x_1^n}\right) dx - \\
&-\int_{\Omega_p} G\varepsilon\left(\frac{\partial^n \boldsymbol{v}}{\partial x_1^n}\right) \varepsilon\left(D_1^{-h}\left(D_1^h \frac{\partial^n \boldsymbol{v}}{\partial x_1^n}\right)\right) dx \quad = \quad -\varrho^s \omega^2 \int_{\Omega_s} \frac{\partial^n \boldsymbol{v}}{\partial x_1^n} \cdot D_1^{-h}\left(D_1^h \frac{\partial^n \boldsymbol{v}}{\partial x_1^n}\right) dx - \\
&\qquad - \varrho^g \omega^2 \int_{\Omega_g} \frac{\partial^n \boldsymbol{v}}{\partial x_1^n} \cdot D_1^{-h}\left(D_1^h \frac{\partial^n \boldsymbol{v}}{\partial x_1^n}\right) dx - \varrho^p \omega^2 \int_{\Omega_p} \frac{\partial^n \boldsymbol{v}}{\partial x_1^n} \cdot D_1^{-h}\left(D_1^h \frac{\partial^n \boldsymbol{v}}{\partial x_1^n}\right) dx.
\end{aligned}$$

We use now that the operation of taking D_1^h is interchangeable with differentiation and

$$\int_U p \cdot D_1^{-h} q \, dx = -\int_U D_1^h p \cdot q \, dx$$

for $U = \Omega_s, \Omega_g, \Omega_p$ and p, $q \in L^2(U)$ such that $p = q = 0$ whenever $x < \frac{\delta}{2} \vee x > l - \frac{\delta}{2}$. Using these properties we can shift D_1^{-h} in every integral of the last integral identity to the other multiplier. We obtain

$$\begin{aligned}
&\mu^s \int\limits_{\Omega_s} \left| \nabla D_1^h \frac{\partial^n \boldsymbol{v}}{\partial x_1^n} \right|^2 dx + (\lambda^s + \mu^s) \int\limits_{\Omega_s} \left(\operatorname{div} D_1^h \frac{\partial^n \boldsymbol{v}}{\partial x_1^n} \right)^2 dx + \\
&+\mu^g \int\limits_{\Omega_g} \left| \nabla D_1^h \frac{\partial^n \boldsymbol{v}}{\partial x_1^n} \right|^2 dx + (\lambda^g + \mu^g) \int\limits_{\Omega_g} \left(\operatorname{div} D_1^h \frac{\partial^n \boldsymbol{v}}{\partial x_1^n} \right)^2 dx + \\
&+ \int\limits_{\Omega_p} G\varepsilon \left(D_1^h \frac{\partial^n \boldsymbol{v}}{\partial x_1^n} \right) \varepsilon \left(D_1^h \frac{\partial^n \boldsymbol{v}}{\partial x_1^n} \right) dx = \\
&= \varrho^s \omega^2 \int\limits_{\Omega_s} \left| D_1^h \frac{\partial^n \boldsymbol{v}}{\partial x_1^n} \right|^2 dx + \varrho^g \omega^2 \int\limits_{\Omega_g} \left| D_1^h \frac{\partial^n \boldsymbol{v}}{\partial x_1^n} \right|^2 dx + \varrho^p \omega^2 \int\limits_{\Omega_p} \left| D_1^h \frac{\partial^n \boldsymbol{v}}{\partial x_1^n} \right|^2 dx.
\end{aligned} \tag{2.55}$$

The last term on the left-hand side (lhs) is treated using the positiveness of G and Korn's inequality (Korn's inequality is applicable because $D_1^h \frac{\partial^n \boldsymbol{v}}{\partial x_1^n}$ is fixed at the side surfaces)

$$\begin{aligned}
\int\limits_{\Omega_p} G\varepsilon \left(D_1^h \frac{\partial^n \boldsymbol{v}}{\partial x_1^n} \right) \varepsilon \left(D_1^h \frac{\partial^n \boldsymbol{v}}{\partial x_1^n} \right) dx \geqslant C \int\limits_{\Omega_p} \left| \varepsilon \left(D_1^h \frac{\partial^n \boldsymbol{v}}{\partial x_1^n} \right) \right|^2 dx &\geqslant C \left\| \nabla D_1^h \frac{\partial^n \boldsymbol{v}}{\partial x_1^n} \right\|^2_{L^2(\Omega_p;\mathbb{R}^{3\times 3})} \\
&= C \left\| D_1^h \nabla \frac{\partial^n \boldsymbol{v}}{\partial x_1^n} \right\|^2_{L^2(\Omega_p;\mathbb{R}^{3\times 3})}.
\end{aligned}$$

Hence for the left-hand side we have

$$\begin{aligned}
(\text{lhs}) \geqslant &\mu^s \left\| D_1^h \nabla \frac{\partial^n \boldsymbol{v}}{\partial x_1^n} \right\|^2_{L^2(\Omega_s;\mathbb{R}^{3\times 3})} + \mu^g \left\| D_1^h \nabla \frac{\partial^n \boldsymbol{v}}{\partial x_1^n} \right\|^2_{L^2(\Omega_g;\mathbb{R}^{3\times 3})} + C \left\| D_1^h \nabla \frac{\partial^n \boldsymbol{v}}{\partial x_1^n} \right\|^2_{L^2(\Omega_p;\mathbb{R}^{3\times 3})} \geqslant \\
\geqslant &\min\{\mu^s, \mu^g, C\} \left\| D_1^h \nabla \frac{\partial^n \boldsymbol{v}}{\partial x_1^n} \right\|^2_{L^2(\Omega_{pgs};\mathbb{R}^{3\times 3})}.
\end{aligned} \tag{2.56}$$

The right-hand side of (2.55) is estimated using (2.46) as follows:

$$\begin{aligned}
(\text{rhs}) &\leqslant \max\{\varrho^s, \varrho^g, \varrho^p\}\, \omega^2 \int\limits_{\Omega_{pgs}} \left| D_1^h \frac{\partial^n \boldsymbol{v}}{\partial x_1^n} \right|^2 dx \leqslant C \int\limits_{\Omega_{pgs}} \left| \frac{\partial^{n+1} \boldsymbol{v}}{\partial x_1^{n+1}} \right|^2 dx = \\
&= C \left\| \frac{\partial^{n+1} \boldsymbol{v}}{\partial x_1^{n+1}} \right\|^2_{L^2(\Omega_{pgs};\mathbb{R}^3)}.
\end{aligned} \tag{2.57}$$

Combining (2.56) and (2.57), we obtain

$$\left\| D_1^h \nabla \frac{\partial^n \boldsymbol{v}}{\partial x_1^n} \right\|^2_{L^2(\Omega_{pgs};\mathbb{R}^3)} \leqslant C \left\| \frac{\partial^{n+1} \boldsymbol{v}}{\partial x_1^{n+1}} \right\|^2_{L^2(\Omega_{pgs};\mathbb{R}^3)},$$

where C is some positive constant that does not depend on h. Obviously the inequality remains valid if we take just one component of the gradient on the left-hand-side:

$$\left.\begin{array}{l} \left\| D_1^h \dfrac{\partial}{\partial x_1} \dfrac{\partial^n \boldsymbol{v}}{\partial x_1^n} \right\|^2_{L^2(\Omega_{pgs};\mathbb{R}^3)} \\ \left\| D_1^h \dfrac{\partial}{\partial x_2} \dfrac{\partial^n \boldsymbol{v}}{\partial x_1^n} \right\|^2_{L^2(\Omega_{pgs};\mathbb{R}^3)} \\ \left\| D_1^h \dfrac{\partial}{\partial x_3} \dfrac{\partial^n \boldsymbol{v}}{\partial x_1^n} \right\|^2_{L^2(\Omega_{pgs};\mathbb{R}^3)} \end{array}\right\} \leqslant C \left\| \frac{\partial^{n+1} \boldsymbol{v}}{\partial x_1^{n+1}} \right\|^2_{L^2(\Omega_{pgs};\mathbb{R}^3)}$$

By Theorem 2.11 this implies that

$$\frac{\partial}{\partial x_1}\left(\frac{\partial^{n+1} \boldsymbol{v}}{\partial x_1^{n+1}}\right), \ \frac{\partial}{\partial x_1}\left(\frac{\partial^{n+1} \boldsymbol{v}}{\partial x_2 \partial x_1^{n}}\right), \ \frac{\partial}{\partial x_1}\left(\frac{\partial^{n+1} \boldsymbol{v}}{\partial x_3 \partial x_1^{n}}\right) \in L^2(\Omega_{pgs};\mathbb{R}^3)$$

and the L^2 norms of these functions are bounded by $C \left\| \dfrac{\partial^{n+1} \boldsymbol{v}}{\partial x_1^{n+1}} \right\|^2_{L^2(\Omega_{pgs};\mathbb{R}^3)}$.

This means $\dfrac{\partial^{n+1} \boldsymbol{v}}{\partial x_1^{n+1}} \in H^1(\Omega_{pgs};\mathbb{R}^3)$ and hence proves (i). Besides, this gives us the estimate

$$\left\| \frac{\partial^{n+2} \boldsymbol{v}}{\partial x_1^{n+2}} \right\|_{L^2(\Omega_{pgs};\mathbb{R}^3)} \leqslant C \left\| \frac{\partial^{n+1} \boldsymbol{v}}{\partial x_1^{n+1}} \right\|_{L^2(\Omega_{pgs};\mathbb{R}^3)}. \tag{2.58}$$

The constant C here is determined by the fixed model parameters $(\mu, \varrho, G, \omega)$, by Korn's constant, and by the constant from (2.46). The last two constants depend only on the domain. Therefore, C does not depend on n and (2.58) holds for any $n \in \mathbb{N}_0$ with the same constant C. We can then apply (2.58) recursively n times to obtain

$$\left\| \frac{\partial^{n+2} \boldsymbol{v}}{\partial x_1^{n+2}} \right\|_{L^2(\Omega_{pgs};\mathbb{R}^3)} \leqslant C^{n+1} \left\| \frac{\partial \boldsymbol{v}}{\partial x_1} \right\|_{L^2(\Omega_{pgs};\mathbb{R}^3)}.$$

The estimate for $\dfrac{\partial \boldsymbol{v}}{\partial x_1}$ is derived in the same way as above with $n = 0$, except that $\boldsymbol{w} := \boldsymbol{v}$

is substituted in (2.54). The arising constant is in generally different from C above because (2.46) is not used. Hence we take the biggest of these two constants and denote it by C. Then we have

$$\left\| \frac{\partial^n \boldsymbol{v}}{\partial x_1^n} \right\|_{L^2(\Omega_{pgs};\mathbb{R}^3)} \leqslant C^n \|\boldsymbol{v}\|_{L^2(\Omega_{pgs};\mathbb{R}^3)} .$$

This is exactly (2.53). Thus (ii) is proved.

The next step is to show that for any $s_2 \in (0, w)$, $s_3 \in (-h_p, h_{gs})$ the function

$$\boldsymbol{u}(x_1) := \int_0^{s_2} \int_{-h_p}^{s_3} \boldsymbol{v}(x_1, x_2, x_3)\, dx_2\, dx_3$$

vanishes for all $x_1 \in (0, l)$. Note that $\boldsymbol{v} \in H^1(\Omega_{pgs}; \mathbb{R}^3)$ implies

$$\boldsymbol{v}\Big|_{x_1=\text{const}} \in H^{1/2}\left(\Omega_{pgs} \cap \{x_1 = \text{const}\}; \mathbb{R}^3\right) .$$

Hence $\boldsymbol{u}(x_1)$ is well-defined everywhere in $(0, l)$. Similarly $\dfrac{\partial^n \boldsymbol{v}}{\partial x_1^n} \in H^1(\Omega_{pgs}; \mathbb{R}^3)$. Therefore,

$$\int_0^{s_2} \int_{-h_p}^{s_3} \left(\frac{\partial^n}{\partial x_1^n} \boldsymbol{v}(x_1, x_2, x_3) \right)^2 dx_2\, dx_3 < \infty \qquad \forall n \in \mathbb{N}_0$$

We can then differentiate $\boldsymbol{u}$ with respect to x_1 and swap the integration and differentiation. This yields

$$\boldsymbol{u}^{(n)}(x_1) := \frac{d^n \boldsymbol{u}(x_1)}{dx_1^n} = \frac{d^n}{dx_1^n} \int_0^{s_2} \int_{-h_p}^{s_3} \boldsymbol{v}(x_1, x_2, x_3)\, dx_2\, dx_3 = \int_0^{s_2} \int_{-h_p}^{s_3} \frac{\partial^n}{\partial x_1^n} \boldsymbol{v}(x_1, x_2, x_3)\, dx_2\, dx_3.$$

Using (2.53), we derive now the following estimate for $\boldsymbol{u}^{(n)}$ in L^2:

$$\left\|\boldsymbol{u}^{(n)}\right\|^2_{L^2(0,\,l)} = \int_0^l \left(\boldsymbol{u}^{(n)}\right)^2 dx_1 = \int_0^l \left(\int_0^{s_2} \int_{-h_p}^{s_3} \frac{\partial^n \boldsymbol{v}}{\partial x_1^n}\, dx_2 dx_3 \right)^2 dx_1 \leqslant$$

$$\leqslant \int_0^l \int_0^{s_2} \int_{-h_p}^{s_3} \left(\frac{\partial^n \boldsymbol{v}}{\partial x_1^n} \right)^2 dx_2 dx_3 dx_1 \leqslant \left\| \frac{\partial^n \boldsymbol{v}}{\partial x_1^n} \right\|^2_{L^2(\Omega_{pgs};\mathbb{R}^3)} \leqslant C^{2n} \|\boldsymbol{v}\|^2_{L^2(\Omega_{pgs};\mathbb{R}^3)} .$$

For the components of $\boldsymbol{u}$ this implies

$$\left\| u_i^{(n)} \right\|_{L^2(0,\,l)} \leqslant C^n \left\| \boldsymbol{v} \right\|_{L^2(\Omega_{pgs};\mathbb{R}^3)}, \qquad i = 1, 2, 3. \tag{2.59}$$

Recall that $\boldsymbol{v}$ vanishes in the damping domain Ω^d_{pgs}, i.e.

$$\boldsymbol{u}(x_1) = 0 = \boldsymbol{v}(x_1,\, x_2,\, x_3) \qquad \text{for } x_1 \in (0,\, \delta) \cup (l - \delta,\, l). \tag{2.60}$$

We apply now Taylor's theorem to $\boldsymbol{u}$ around the point $x_1 = \delta/2$ with the remainder term in the integral form:

$$\boldsymbol{u}(x_1) = \sum_{k=0}^{n} \frac{\boldsymbol{u}^{(k)}(\delta/2)}{k!} \left(x_1 - \frac{\delta}{2} \right)^k + \int\limits_{\delta/2}^{x_1} \frac{(x_1 - t)^n}{n!} \boldsymbol{u}^{(n+1)}(t) dt, \qquad \forall x_1 \in \left[\frac{\delta}{2},\, l \right).$$

Due to (2.60) $\boldsymbol{u}^{(n)}(\delta/2) = 0$ for all $n \in \mathbb{N}_0$. We have then

$$|u_i(x_1)| = \left| \int\limits_{\delta/2}^{x_1} \frac{(x_1 - t)^n}{n!} u_i^{(n+1)}(t) dt \right| \leqslant \frac{1}{n!} \left(\int\limits_0^l (x_1 - t)^{2n} dt \right)^{\frac{1}{2}} \left(\int\limits_0^l \left(u_i^{(n+1)}(t) \right)^2 dt \right)^{\frac{1}{2}} =$$

$$= \frac{1}{n!} \left(\frac{(l - x_1)^{2n+1} + x_1^{2n+1}}{2n+1} \right)^{\frac{1}{2}} \left\| u_i^{(n+1)} \right\|_{L^2(0,\,l)} \leqslant \sqrt{2l} \frac{l^n C^{n+1}}{n!(2n+1)^{\frac{1}{2}}} \left\| \boldsymbol{v} \right\|_{L^2(\Omega_{pgs};\mathbb{R}^3)}.$$

We used here Hölder's inequality and (2.59). The inequality above holds for any $n \in \mathbb{N}$ and the term on the right-hand side goes to zero as $n \to \infty$. Therefore,

$$\boldsymbol{u}(x_1) = \int\limits_0^{s_2} \int\limits_{-h_p}^{s_3} \boldsymbol{v}(x_1, x_2, x_3)\, dx_2\, dx_3 = 0 \qquad \forall\, x_1 \in (0,\, l).$$

Recall now that s_2 and s_3 were taken arbitrary from the intervals $(0,\, w)$ and $(-h_p,\, h_{gs})$ respectively. Hence,

$$\int\limits_{t_2}^{s_2} \int\limits_{t_3}^{s_3} \boldsymbol{v}(x_1, x_2, x_3)\, dx_2\, dx_3 = 0 \qquad \forall\, x_1 \in (0,\, l)$$

for any rectangular neighborhood $(t_2,\, s_3) \times (t_3,\, s_3) \subset (0,\, w) \times (-h_p,\, h_{gs})$. This means that for every fixed $x_1 \in (0, l)$

$$\boldsymbol{v}(x_1, x_2, x_3) = 0 \quad \text{a.e. in } (0, w) \times (-h_p, h_{gs})$$

and therefore $\boldsymbol{v} = 0$ a.e. in Ω_{pgs}. Thus $u = (\boldsymbol{v}^{(1)}, \boldsymbol{v}^{(2)}, \varphi^{(1)}, \varphi^{(2)}) = 0$.

□

Combining Theorem 2.7, Lemma 2.8, Theorem 2.10 and Theorem 2.12, we obtain the well-posedness result for our model.

Theorem 2.13. *Problem 2.2 possesses a unique solution that depends continuously on the functional on the right-hand side.*

2.6 Numerical Treatment

2.6.1 Ritz-Galerkin Approximation

In this section we describe the discretization of the model, discuss the solvability of the discretized problem and the convergence of its solution to the solution of the original problem.

Denote by $\{\mathcal{V}^N\}_{N \in \mathbb{N}}$ a sequence of finite-dimensional spaces such that

$$\forall N \in \mathbb{N} \quad \mathcal{V}^N \subset \mathcal{V}, \tag{2.61}$$

$$\forall u \in \mathcal{V} \quad \inf_{v \in \mathcal{V}^N} \|u - v\|_{\mathcal{V}} \to 0 \quad \text{as } N \to \infty. \tag{2.62}$$

Note that $\mathcal{V}$ is a separable space by construction and hence such a sequence always exists. Further, denote by $S_{\mathcal{V}^N}$ the unit spheres in $\mathcal{V}^N$, i.e.

$$S_{\mathcal{V}^N} := \{v \in \mathcal{V}^N : \|v\|_{\mathcal{V}} = 1\}.$$

We assume that $\mathcal{V}^N$ is equiped with the norm $\|\cdot\|_{\mathcal{V}}$.

Remark 2.14. The infimum in (2.62) is reached for all fixed $N \in \mathbb{N}, u \in \mathcal{V}$ and therefore can be replaced by the minimum. Indeed, suppose $\{v^m\} \subset \mathcal{V}^N$ is a minimizing sequence of $\|u - v\|_{\mathcal{V}}$ for some fixed u. Then we have

$$\|v^m\|_{\mathcal{V}} \leqslant \|v^m - u\|_{\mathcal{V}} + \|u\|_{\mathcal{V}} \to \inf_{v \in \mathcal{V}^N} \|u - v\|_{\mathcal{V}} + \|u\|_{\mathcal{V}} = C < \infty.$$

This means that v^m is bounded. Since $\mathcal{V}^N$ is a finite-dimensional space, the boundedness implies the precompactness. Therefore there exist a subsequence of $\{v^m\}$, that we denote by the same index m, and $v^N \in \mathcal{V}^N$ such that $v^m \to v^N$ in $\mathcal{V}^N$ (and consequently in $\mathcal{V}$). Then by the triangle inequality

$$\left.\begin{array}{l} \|u - v^m\|_{\mathcal{V}} - \|u - v^N\|_{\mathcal{V}} \leqslant \|v^m - v^N\|_{\mathcal{V}} \\ \|u - v^N\|_{\mathcal{V}} - \|u - v^m\|_{\mathcal{V}} \leqslant \|v^m - v^N\|_{\mathcal{V}} \end{array}\right\} \Rightarrow \left|\|u - v^m\|_{\mathcal{V}} - \|u - v^N\|_{\mathcal{V}}\right| \leqslant \|v^m - v^N\|_{\mathcal{V}}.$$

Taking the limit as $m \to \infty$, we obtain $\|u - v^m\|_{\mathcal{V}} \to \|u - v^N\|_{\mathcal{V}}$, which means that v^N is a minimizer.

Remark 2.15. Suppose $u \in S_{\mathcal{V}}$. Then (2.62) implies

$$\inf_{v \in S_{\mathcal{V}^N}} \|u - v\|_{\mathcal{V}} \to 0 \quad \text{as } N \to \infty. \tag{2.63}$$

Indeed, let $v^N \in \mathcal{V}^N$ be a sequence of minimizers of $\|u - v\|_{\mathcal{V}}$ over $v \in \mathcal{V}^N$. They exist due to Remark 2.14. According to (2.62) $\|u - v^N\|_{\mathcal{V}} \to 0$ as $N \to \infty$. This implies $\|v^N\|_{\mathcal{V}} \to \|u\|_{\mathcal{V}} = 1$. Let us define

$$w^N := \frac{v^N}{\|v^N\|}.$$

Note that $w^N \in S_{\mathcal{V}^N}$. We have then

$$\begin{aligned} 0 \leqslant \inf_{v \in S_{\mathcal{V}^N}} \|u - v\|_{\mathcal{V}} \leqslant \|u - w^N\|_{\mathcal{V}} \leqslant \\ \leqslant \|u - v^N\|_{\mathcal{V}} + \|v^N - w^N\|_{\mathcal{V}} = \\ = \|u - v^N\|_{\mathcal{V}} + \left|1 - \frac{1}{\|v^N\|}\right| \|v^N\|_{\mathcal{V}}. \end{aligned}$$

Passing here N to ∞, we obtain (2.63).

We pose now a discrete counterpart of Problem 2.2.

Problem 2.4. *Find $u^N \in \mathcal{V}^N$ such that*

$$\pi(u^N, v) = \ell(v) \quad \forall v \in \mathcal{V}^N.$$

The solution of Problem 2.4, if it exists, is called the *Ritz-Galerkin solution* of the original problem (Problem 2.2).

The existence of the Ritz-Galerkin solution and its convergence to the solution of the exact problem is based on the following theorem.

Theorem 2.16. *Let $\{\mathcal{V}^N\}_{N\in\mathbb{N}}$ be a sequence of subspaces defined as above, $\pi(\cdot,\cdot)$ - a bilinear form bounded on $\mathcal{V} \times \mathcal{V}$. For each $N \in \mathbb{N}$ define*

$$\epsilon^N := \inf_{u\in S_{\mathcal{V}^N}} \sup_{v\in S_{\mathcal{V}^N}} \pi(u,v).$$

Obviously $\epsilon^N \geqslant 0$.
Suppose that there is $\tilde{\epsilon} \in \mathbb{R}$ such that

$$\epsilon_N \geqslant \tilde{\epsilon} > 0 \quad \textit{for all} \ \ N \in \mathbb{N}.$$

Then Problem 2.4 is uniquely solvable and the Ritz-Galerkin solutions u^N converge to u, i.e.

$$\|u - u^N\|_{\mathcal{V}} \to 0 \quad \textit{as } N \to \infty.$$

The last statement follows directly from the quasi-best approximation property that was first proved by Ivo Babuška in [5]. The existence and uniqueness of Ritz-Galerkin solutions are due to Theorem 2.7.

Theorem 2.17. *Let $\mathcal{W}, \mathcal{V}$, $\{\mathcal{V}^N\}$ and $\pi(\cdot,\cdot)$ be defined as above. Then there exist $M \in \mathbb{N}$ and $\tilde{\epsilon} \in \mathbb{R}$ such that*

$$\epsilon^N := \inf_{u\in S_{\mathcal{V}^N}} \sup_{v\in S_{\mathcal{V}^N}} \pi(u,v) \geqslant \tilde{\epsilon} > 0 \qquad \forall N > M. \tag{2.64}$$

Proof. Suppose the claim is false. Then there exist sequences $\mu^m \to 0$ and $u^m \in S_{\mathcal{V}^m}$ such that

$$\sup_{v\in S_{\mathcal{V}^m}} \pi(u^m, v) < \mu^m.$$

Arguing as in the proof of Theorem 2.10 we can construct a sequence $\{w^m\} \subset \mathcal{V}$ such that:

- w_m is bounded in $\mathcal{V}$,
- $w^m \rightharpoonup w^0 \neq 0$ in $\mathcal{V}$ and consequently $w^m \to w^0$ in $\mathcal{W}$,
- the following inequality holds:

$$\sup_{v \in S_{\mathcal{V}^m}} \pi(w^m, v) < \mu^m \|w^m\|_{\mathcal{V}}. \tag{2.65}$$

Furthermore, in the proof of Theorem 2.10, we showed that there exist $v^0 \in \mathcal{V}$ (but not necessarily in $S_{\mathcal{V}}$) such that

$$\pi(w^0, v^0) = \sup_{v \in S_{\mathcal{V}}} \pi(w^0, v).$$

Suppose $v^0 = 0$. Then

$$0 \leqslant \inf_{u \in S_{\mathcal{V}}} \sup_{v \in S_{\mathcal{V}}} \pi(u, v) \leqslant \sup_{v \in S_{\mathcal{V}}} \pi\left(\frac{w^0}{\|w^0\|_{\mathcal{V}}}, v\right) = \frac{1}{\|w^0\|_{\mathcal{V}}} \pi(w^0, v^0) = 0.$$

Consequently,

$$\inf_{u \in S_{\mathcal{V}}} \sup_{v \in S_{\mathcal{V}}} \pi(u, v) = 0.$$

On the other hand, in Section 2.5 we showed (see Theorems 2.10 and 2.12) that

$$\inf_{u \in S_{\mathcal{V}}} \sup_{v \in S_{\mathcal{V}}} \pi(u, v) > 0. \tag{2.66}$$

Hence $v^0 \neq 0$. We can then define

$$\tilde{v}^0 := \frac{v^0}{\|v^0\|_{\mathcal{V}}} \in S_{\mathcal{V}}.$$

Then the following estimate holds:

$$\begin{aligned}
\mu^m \|w_m\|_{\mathcal{V}} > & \sup_{v \in S_{\mathcal{V}^m}} \pi(w^m, v) \\
& = \pi(w^m, \tilde{v}^0) + \sup_{v \in S_{\mathcal{V}^m}} \pi(w^m, v - \tilde{v}^0) = \\
& = \pi(w^m, \tilde{v}^0) - \inf_{v \in S_{\mathcal{V}^m}} \pi(w^m, \tilde{v}^0 - v) \geqslant \\
& \geqslant \pi(w^m, \tilde{v}^0) - C\|w^m\|_{\mathcal{V}} \inf_{v \in S_{\mathcal{V}^m}} \|\tilde{v}^0 - v\|_{\mathcal{V}}.
\end{aligned}$$

We used here (2.65) and the boundedness of π. Recall that w^m is bounded in $\mathcal{V}$, $w^m \rightharpoonup w^0$ and $\inf\limits_{v \in S_{\mathcal{V}^m}} \|\tilde{v}^0 - v\|_{\mathcal{V}} \to 0$ as $m \to \infty$ by Remark 2.15. Hence taking the limit as $m \to \infty$ yields

$$\pi(w^0, \tilde{v}^0) \leqslant 0.$$

By construction of $\tilde{v}^0$ this implies

$$\sup_{v \in S_{\mathcal{V}}} \pi(w^0, v) = 0.$$

Therefore,

$$\inf_{u \in S_{\mathcal{V}}} \sup_{v \in S_{\mathcal{V}}} \pi(u, v) = 0.$$

This contradicts (2.66).

□

Combining Theorems 2.16 and 2.17, we obtain the convergence of the Ritz-Galerkin solutions to the solution of the original problem. The existence and uniqueness of the Ritz-Galerkin solution take place when N is big enough.

2.6.2 Domain Decomposition

The convergence of the Ritz-Galerkin to the exact solution of the problem established in the previous subsection enables us to use the finite element method (FEM) for the simulation of the sensor. However the straightforward serial calculation by this method turns to be very time- and resource-consuming. The main reason for this is the smallness of the wavelength in comparison with the size of the sensor. For example, the wavelength at the typical operating frequency 100 MHz is about 45μm, while the x_1-length of the sensor, i.e. the way to travel, is at least 1.8 mm. Taking 8 gridpoints per wavelength we arrive at 320 divisions in x_1-direction only. Though the discretization in x_2- and x_3-directions does not have to be that fine, the overall number of degrees of freedom is still significant due to the fact that we have 8 unknown scalar functions. For example, the mesh $320 \times 50 \times 20$ yields 2 560 000 degrees of freedom. Such an amount encourages us to develop a parallel implementation of the model.

One way to bring parallelism into the implementation is to exploit parallel sparse solvers when applying the finite element method. We used two of them - `SPOOLES` and `PARDISO`.

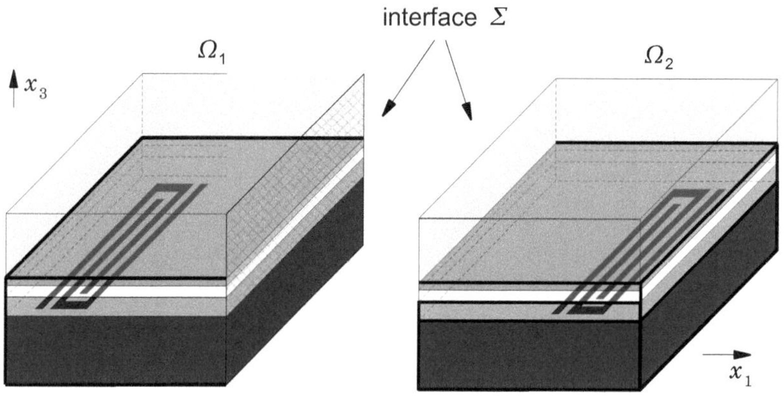

Figure 2.4: Domain decomposition.

PARDISO provides an efficient way to solve sparse systems with symmetric multiprocessing (SMP). SPOOLES is additionally available in a version optimized for calculation with MPI, which makes it suitable for using on computer clusters. More details about these two libraries can be found in [8] and [61]. Parallel sparse solvers bring more efficiency only at the stage of the solving the sparse linear systems. They do not require any significant change in the model to be applicable.

Another approach to parallelization is based on domain decomposition and requires some more analysis. It can also be applied together with parallel sparse solvers. This subsection is devoted to the description and mathematical foundation of this method.

The main idea is to split the original domain into several non-overlaping subdomains, prescribe natural boundary conditions on the interfaces between the subdomains, solve the problem in each domain independently, and then iteratively adjust the natural conditions minimizing the discontinuity in the solution at the interfaces. For the sake of simplicity we consider just two subdomains.

We note here that this approach may be significantly slower than a serial calculation and the main reason for using it is not the speedup but the ability to handle bigger problems due to the spreading the mesh through the computational nodes.

Let Ω_1 and Ω_2 be two open subdomains of Ω resulted from splitting Ω by some plane orthogonal to x_1, i.e

$$\Omega = \Omega_1 \cup \Sigma \cup \Omega_2,$$

where Σ is the interface between the subdomains (see Figure 2.4). Recall that by construction we can consider $\mathcal{V}$ as a Hilbert space of vector-valued H^1-functions defined on Ω and Ω_p, i.e.

$$\begin{aligned}\mathcal{V} = \{(\boldsymbol{v}, \boldsymbol{w}, \varphi, \psi) \in \left[H^1(\Omega;\mathbb{R}^3)\right]^2 \oplus \left[H^1(\Omega_p)\right]^2 : \boldsymbol{v}|_{\Gamma_5} = \boldsymbol{w}|_{\Gamma_5} = 0, \\ \varphi|_{S_1\cup S_2\cup S_3} = \psi|_{S_1\cup S_2\cup S_3} = 0, \\ \varphi|_{S_4} = \text{const}, \ \psi|_{S_4} = \text{const}\}\end{aligned}$$

Let us construct the Hilbert spaces $\mathcal{V}_1$ and $\mathcal{V}_2$ by restricting the functions from $\mathcal{V}$ on Ω_1 and Ω_2 respectively:

$$\mathcal{V}_1 := \{u \in \left[H^1(\Omega_1;\mathbb{R}^3)\right]^2 \oplus \left[H^1(\Omega_p \cap \Omega_1)\right]^2 : \exists v \in \mathcal{V} \text{ such that } v\Big|_{\Omega_1} = u\},$$

$$\mathcal{V}_2 := \{u \in \left[H^1(\Omega_2;\mathbb{R}^3)\right]^2 \oplus \left[H^1(\Omega_p \cap \Omega_2)\right]^2 : \exists v \in \mathcal{V} \text{ such that } v\Big|_{\Omega_2} = u\}.$$

The norm and the inner product in $\mathcal{V}_1$ and $\mathcal{V}_2$ are defined the same way as in $\mathcal{V}$ except that the integration is performed over Ω_1 and Ω_2 respectively.

Similarly, let us define the bilinear forms $\pi_1(\cdot,\cdot)$ on $\Omega_1 \times \Omega_1$ and $\pi_2(\cdot,\cdot)$ on $\Omega_2 \times \Omega_2$. They are constructed the same way as $\pi(\cdot,\cdot)$ on the base of the integral identity (2.33), except that the integrations in (2.33) are performed over subdomains of Ω intersected with Ω_1 and Ω_2 respectively. Obviously $\pi_1(\cdot,\cdot)$ and $\pi_2(\cdot,\cdot)$ possess the same properties as $\pi(\cdot,\cdot)$ (see Proposition 2.5). Denote by A_1 and A_2 the operators associated to $\pi_1(\cdot,\cdot)$ and $\pi_1(\cdot,\cdot)$ respectively.

Further, since elements of $\mathcal{V}_1$ and $\mathcal{V}_2$ are composed of H^1-functions, we can consider their traces on Σ. Define the trace operators as follows:

$$\begin{aligned}B_1 : \mathcal{V}_1 \ni (\boldsymbol{v}, \boldsymbol{w}, \varphi, \psi) \mapsto \left(\boldsymbol{v}\Big|_{\Sigma}, \boldsymbol{w}\Big|_{\Sigma}, \varphi\Big|_{\Omega_p\cap\Sigma}, \psi\Big|_{\Omega_p\cap\Sigma}\right), \\ B_2 : \mathcal{V}_2 \ni (\boldsymbol{v}, \boldsymbol{w}, \varphi, \psi) \mapsto \left(\boldsymbol{v}\Big|_{\Sigma}, \boldsymbol{w}\Big|_{\Sigma}, \varphi\Big|_{\Omega_p\cap\Sigma}, \psi\Big|_{\Omega_p\cap\Sigma}\right).\end{aligned}$$

The image of $B_1(\mathcal{V}_1)$ and $B_2(\mathcal{V}_2)$ is

$$\mathcal{S} := \{(\boldsymbol{v}, \boldsymbol{w}, \varphi, \psi) \in \left[H^{1/2}(\Sigma;\mathbb{R}^3)\right]^2 \oplus \left[H^{1/2}(\Omega_p \cap \Sigma)\right]^2 : \boldsymbol{v}\Big|_{\Gamma_5\cap\Sigma} = \boldsymbol{w}\Big|_{\Gamma_5\cap\Sigma} = 0\}.$$

This is a Hilbert space with the inner product induced by that of $H^{1/2}$. Every functional from $\mathcal{S}'$ can be identified with some natural boundary condition on Σ. From the physical point of view it prescribes the normal pressure and the normal electrical displacements on the interface.

We are ready now to formulate the problems for the subdomains.

Problem 2.5. *Let $s \in \mathcal{S}'$. Find $u_1 \in \mathcal{V}_1$ such that*

$$\pi_1(u_1, v) = \ell v + sB_1 v, \quad \forall\, v \in \mathcal{V}_1,$$

or, equivalently,

$$A_1 u_1 = \ell + s_\circ B_1.$$

The functional ℓ here is the same ℓ defined for the right-hand side of the integral identity (2.33). Originally it was defined as a functional on $\mathcal{V}$ not on $\mathcal{V}_1$ and represented the electrical boundary condition on the input electrodes. But actually ℓ is a L^2-function with the support lying close around the input electrodes. We assume here that the interface Σ cuts the biosensor far enough from the input electrodes so that the Ω_2-part of $\mathcal{V}$-functions does not influence the value of ℓ. Hence, we can consider ℓ as a functional on $\mathcal{V}_1$.

Problem 2.6. *Let $s \in \mathcal{S}'$. Find $u_2 \in \mathcal{V}_2$ such that*

$$\pi_2(u_2, v) = -sB_2 v, \quad \forall\, v \in \mathcal{V}_2.$$

or, equivalently,

$$A_2 u_2 = -s_\circ B_2.$$

The following theorem establishes the well-posedness of the problems.

Theorem 2.18. *For all $s \in \mathcal{S}'$ Problems 2.5 and 2.6 possess unique solutions that depend continuously on the functional on the right-hand side, i.e. the operators A_1^{-1} and A_2^{-1} exist and they are bounded.*

Proof. Since $\pi_1(\cdot,\cdot)$ and $\pi_2(\cdot,\cdot)$ are of the same structure as $\pi(\cdot,\cdot)$, the proof can be performed the same way as for Problem 2.2 in Section 2.5. The only difference is that when proving the uniqueness of the trivial solution of the homogeneous problem (see Theorem 2.12) the difference quotient is to be taken not in the x_1- but in the x_2-direction.

□

Since A_1^{-1} and A_2^{-1} are well-defined, we can consider solutions u_1 and u_2 as functions of s, i.e.

$$\begin{aligned} u_1(s) &= A_1^{-1}(\ell + s_\circ B_1), \\ u_2(s) &= A_2^{-1}(-s_\circ B_2). \end{aligned} \tag{2.67}$$

Theorem 2.19. *Suppose there exists $s \in \mathcal{S}'$ such that the solutions $u_1(s)$ and $u_2(s)$ share the same trace on Σ, i.e.*

$$B_1 u_1(s) = B_2 u_2(s). \tag{2.68}$$

Then the function

$$u(\boldsymbol{x}) := \begin{cases} u_1(s)(\boldsymbol{x}), & \text{if } \boldsymbol{x} \in \Omega_1, \\ u_2(s)(\boldsymbol{x}), & \text{if } \boldsymbol{x} \in \Omega_2 \end{cases}$$

belongs to $\mathcal{V}$ and solves Problem 2.2, i.e.

$$\pi(u, v) = \ell v, \quad \forall\, v \in \mathcal{V}. \tag{2.69}$$

Proof. We first prove that (2.68) implies that u is a solution of Problem 2.2. Let u be of the form $(\boldsymbol{v}, \boldsymbol{w}, \varphi, \psi)$ composed from $u_1 = (\boldsymbol{v}_1, \boldsymbol{w}_1, \varphi_1, \psi_1)$ and $u_2 = (\boldsymbol{v}_2, \boldsymbol{w}_2, \varphi_2, \psi_2)$. To prove that $u \in \mathcal{V}$ it is sufficient to show that all the components of u are H^1-functions. Let us show that $\varphi \in H^1(\Omega_p)$. For $i = 1, 2, 3$ define

$$\partial_i \varphi(\boldsymbol{x}) := \begin{cases} \dfrac{\partial \varphi_1}{\partial x_i}(\boldsymbol{x}), & \text{if } \boldsymbol{x} \in \Omega_1, \\ \dfrac{\partial \varphi_2}{\partial x_i}(\boldsymbol{x}), & \text{if } \boldsymbol{x} \in \Omega_2 \end{cases}$$

Obviously $\partial_i \varphi \in L^2(\Omega_p)$. For all $g \in C_0^\infty(\Omega_p)$ holds:

$$\begin{aligned} -\int_{\Omega_p} \varphi \frac{\partial g}{\partial x_i} dx &= -\int_{\Omega_p \cap \Omega_1} \varphi_1 \frac{\partial g}{\partial x_i} dx - \int_{\Omega_p \cap \Omega_2} \varphi_2 \frac{\partial g}{\partial x_i} dx = \\ &= \int_{\Omega_p \cap \Omega_1} \frac{\partial \varphi_1}{\partial x_i} g\, dx + \int_{\Omega_p \cap \Omega_2} \frac{\partial \varphi_2}{\partial x_i} g\, dx - \int_{\Omega_p \cap \Sigma} \left(\varphi_1\big|_\Sigma - \varphi_2\big|_\Sigma \right) g\big|_\Sigma\, n_i\, dx = \\ &= \int_{\Omega_p} \partial_i \varphi\, g\, dx, \end{aligned}$$

where n_i is the i-th component of the unit normal vector pointing from Ω_1 to Ω_2. Here we used the integration by parts and the fact that $\varphi_1\big|_\Sigma = \varphi_2\big|_\Sigma$ as follows from (2.68). The equality above means that $\partial_i \varphi$ is the weak derivative of φ and therefore $\varphi \in H^1(\Omega_p)$.

In the same way it can be shown that other components of u are weak differentiable and hence $u \in \mathcal{V}$. We can then substitute it in $\pi(\cdot,\cdot)$. By construction of π_1 and π_2 for all $v \in \mathcal{V}$ holds:

$$\begin{aligned}\pi(u,v) &= \pi_1(u\big|_{\Omega_1}, v\big|_{\Omega_1}) + \pi_2(u\big|_{\Omega_2}, v\big|_{\Omega_2}) = \\ &= \ell v + sB_1 v\big|_{\Omega_1} - sB_2 v\big|_{\Omega_2} = \\ &= \ell v\end{aligned}$$

and therefore u satisfies (2.69). □

Remark. Let $u = (\boldsymbol{v}^{(1)}, \boldsymbol{v}^{(2)}, \varphi^{(1)}, \varphi^{(2)})$ be the solution of Problem 2.2. Assume that the components of u are H^2-functions. In this case the stress tensor $\sigma(u)$ and electrical displacements $\boldsymbol{D}(u)$ belong to H^1 and the functional s satisfying (2.68) can be constructed explicitely as follows

$$\begin{aligned} s:\ v = (\boldsymbol{v}^{(1)}, \boldsymbol{v}^{(2)}, \psi^{(1)}, \psi^{(2)}) \ \mapsto\ & \int_{\Sigma} (\sigma(u)\cdot\boldsymbol{n})\cdot\boldsymbol{v}^{(1)} ds + \int_{\Sigma} (\sigma(u)\cdot\boldsymbol{n})\cdot\boldsymbol{v}^{(2)} ds + \\ & + \int_{\Sigma\cap\Omega_p} (\boldsymbol{D}(u)\cdot\boldsymbol{n})\,\psi^{(1)} ds + \int_{\Sigma\cap\Omega_p} (\boldsymbol{D}(u)\cdot\boldsymbol{n})\,\psi^{(2)} ds. \end{aligned} \tag{2.70}$$

For this s $u\big|_{\Omega_1}$ solves Problem 2.5 and $u\big|_{\Omega_2}$ solves Problem 2.6. This can easily be shown by integrating the corresponding integral identity by parts and using Proposition 2.2. In the general case, though, the definition (2.70) is not correct because σ and $\boldsymbol{D}$ are L^2-functions. They do not have to be weak differentiable and their traces on Σ are in general not defined.

From the physical point of view the functional s describes the influence of the right part of the biosensor on the left part through the surface Σ. This influence is not known a priori, but we can solve Problems 2.5 and 2.6 for some s and estimate its quality by the jump of the solution on the interface, i.e. by the difference $B_1 u_1(s) - B_2 u_2(s)$. This leads to the following optimization problem:

Problem 2.7. *Find*

$$\inf_{u_1\in\mathcal{V}_1, u_2\in\mathcal{V}_2} J(u_1,u_2) := \frac{1}{2}\left\|B_1 u_1 - B_2 u_2\right\|_{\mathcal{S}}^2 \tag{2.71}$$

subject to

$$\begin{aligned} A_1 u_1 &= \ell + s_\circ B_1, \\ A_2 u_2 &= - s_\circ B_2, \quad s \in \mathcal{S}'. \end{aligned}$$

As noted above the operators A_1 and A_2 are invertible and u_1 and u_2 can be considered as functions of s. Inserting $u_1(s)$ and $u_2(s)$ from (2.67) in (2.71) we obtain the reduced problem as follows:

Problem 2.8. *Find*

$$\inf_{s\in\mathcal{S}'} \hat{J}(s) := J(u_1(s), u_2(s)).$$

This is an unconstrained optimization problem now. We solve it iteratively by the well-known conjugate gradient method.

Algorithm 2.1 (Conjugate gradient method).

$i := 0$

// Set the initial descent direction

$d_0 := -\hat{J}'(s_0)$

`repeat`

// Do the line search in the descent direction d_i by the Secant method

$$\alpha_i = -\frac{\langle \hat{J}'(s_i), d_i\rangle_S}{\langle \hat{J}'(d_i), d_i\rangle_S} \quad \left(= -\frac{\langle \hat{J}'(s_i), d_i\rangle_S}{\langle \hat{J}'(s_i + d_i), d_i\rangle_S - \langle \hat{J}'(s_i), d_i\rangle_S} \right)$$

// Move to the next point

$s_{i+1} := s_i + \alpha_i d_i$

// Calculate the next descent direction

$r_{i+1} := -\hat{J}'(s_{i+1})$

$\beta_{i+1} := \dfrac{\langle r_{i+1}, r_{i+1}\rangle_S}{\langle r_i, r_i\rangle_S}$

$d_{i+1} := r_{i+1} + \beta_{i+1} d_i$

$i := i + 1$

`until` r_{i+1} is sufficiently small.

Here $\hat{J}'(s)$ denotes the Fréchet derivative of $\hat{J}(s)$ at the point $s \in \mathcal{S}'$. Since $\mathcal{S}$ is reflexive (moreover, it is a Hilbert space), we can consider $\hat{J}'(s)$ as an element of $\mathcal{S}$. The following theorem establishes the way to finding it.

Theorem 2.20. *The functional $\hat{J}(s)$ is Fréchet differentiable in $\mathcal{S}'$. Its derivative as an element of $\mathcal{S}$ can be found by*

$$\hat{J}'(s) = B_1 p_1(s) - B_2 p_2(s),$$

where $p_1(s) \in \mathcal{V}_1$ and $p_2(s) \in \mathcal{V}_2$ are the unique solutions of the adjoint equations

$$\pi_1^*(p_1(s),\, v) = \ \ \langle B_1 u_1(s) - B_2 u_2(s),\, B_1 v\rangle_S \qquad \forall\, v \in \mathcal{V}_1, \tag{2.72}$$
$$\pi_2^*(p_2(s),\, v) = -\langle B_1 u_1(s) - B_2 u_2(s),\, B_2 v\rangle_S \qquad \forall\, v \in \mathcal{V}_2. \tag{2.73}$$

Here $\pi_1^*(\cdot,\cdot)$ and $\pi_2^*(\cdot,\cdot)$ are the bilinear forms adjoint to $\pi_1(\cdot,\cdot)$ and $\pi_2(\cdot,\cdot)$ respectively. By definition,

$$\pi_1^*(u,\, v) := \pi_1(v,\, u) \quad \forall\, u, v \in \mathcal{V}_1,$$
$$\pi_2^*(u,\, v) := \pi_2(v,\, u) \quad \forall\, u, v \in \mathcal{V}_2.$$

Proof. Let us first establish the differentiability of $\hat{J}(s) = J(u_1(s), u_2(s))$. The differentiability of $J(u_1, u_2)$ with respect to u_1 and u_2 is easily seen from the definition of J (see (2.71)). The derivatives satisfy

$$\begin{aligned} J'_{u_1}(u_1(s), u_2(s))v &= \ \ \langle B_1 u_1(s) - B_2 u_2(s),\, B_1 v\rangle_S, \quad \forall\, v \in \mathcal{V}_1,\\ J'_{u_2}(u_1(s), u_2(s))v &= -\langle B_1 u_2(s) - B_2 u_2(s),\, B_2 v\rangle_S, \quad \forall\, v \in \mathcal{V}_2. \end{aligned} \tag{2.74}$$

Further, the operators A_1^{-1} and A_2^{-1} are linear and bounded, and therefore $u_1(s)$ and $u_2(s)$ are Fréchet differentiable with respect to s. From (2.67) we derive

$$u_1'(s)h = \ \ A_1^{-1}(h \circ B_1) \quad \forall\, h \in \mathcal{S}',$$
$$u_2'(s)h = -A_2^{-1}(h \circ B_2) \quad \forall\, h \in \mathcal{S}'.$$

Then by the chain rule $\hat{J}(s)$ is also Fréchet differentiable, and for all $h \in \mathcal{S}'$ we have

$$\begin{aligned} \langle \hat{J}'(s), h\rangle_{\mathcal{S}'',\mathcal{S}'} &= \langle J'_{u_1}(u_1(s), u_2(s)),\quad u_1'(s)h\ \rangle_{\mathcal{V}_1',\mathcal{V}_1} + \langle J'_{u_2}(u_1(s), u_2(s)),\quad u_2'(s)h\ \rangle_{\mathcal{V}_2',\mathcal{V}_2} =\\ &= \langle J'_{u_1}(u_1(s), u_2(s)), A_1^{-1}(h \circ B_1)\rangle_{\mathcal{V}_1',\mathcal{V}_1} - \langle J'_{u_2}(u_1(s), u_2(s)), A_2^{-1}(h \circ B_2)\rangle_{\mathcal{V}_2',\mathcal{V}_2} =\\ &= \langle A_1^{-*} J'_{u_1}(u_1(s), u_2(s)),\ h \circ B_1\ \rangle_{\mathcal{V}_1'',\mathcal{V}_1'} - \langle A_2^{-*} J'_{u_2}(u_1(s), u_2(s)),\ h \circ B_2\ \rangle_{\mathcal{V}_2'',\mathcal{V}_2'}. \end{aligned}$$

Here $A_1^{-*} \in \mathcal{L}(\mathcal{V}_1', \mathcal{V}_1'')$ and $A_2^{-*} \in \mathcal{L}(\mathcal{V}_2', \mathcal{V}_2'')$ are operators adjoint to A_1^{-1} and A_2^{-1} respectively. Introducing

$$p_1(s) := A_1^{-*} J'_{u_1}(u_1(s), u_2(s)),$$
$$p_2(s) := A_2^{-*} J'_{u_2}(u_1(s), u_2(s)) \tag{2.75}$$

and identifying $\mathcal{V}_1''$ with $\mathcal{V}_1$ and $\mathcal{V}_2''$ with $\mathcal{V}_2$, we obtain

$$\begin{aligned}\langle \hat{J}'(s), h\rangle_{\mathcal{S}'',\mathcal{S}'} &= \langle p_1(s), h\circ B_1\rangle_{\mathcal{V}_1'',\mathcal{V}_1'} - \langle p_2(s), h\circ B_2\rangle_{\mathcal{V}_2'',\mathcal{V}_2'} = \\ &= \langle h\circ B_1, p_1(s)\rangle_{\mathcal{V}_1',\mathcal{V}_1} - \langle h\circ B_2, p_2(s)\rangle_{\mathcal{V}_2',\mathcal{V}_2} = \\ &= \langle h, B_1 p_1(s)\rangle_{\mathcal{S}',\mathcal{S}} - \langle h, B_2 p_2(s)\rangle_{\mathcal{S}',\mathcal{S}} = \\ &= \langle h, B_1 p_1(s) - B_2 p_2(s)\rangle_{\mathcal{S}',\mathcal{S}}.\end{aligned}$$

This means

$$\hat{J}'(s) = B_1 p_1(s) - B_2 p_2(s).$$

It is left to show that $p_1(s)$ and $p_2(s)$ defined by (2.75) and considered as elements of $\mathcal{V}_1$ and V_2 solve (2.72) and (2.73) respectively. Indeed, (2.75) implies

$$A_1^* p_1(s) = J'_{u_1}(u_1(s), u_2(s)),$$
$$A_2^* p_2(s) = J'_{u_2}(u_1(s), u_2(s)).$$

We used here that $(A_i^*)^{-1} = (A_i^{-1})^*, i = 1, 2$. The relations above equate the functionals on $\mathcal{V}_1$ and $\mathcal{V}_2$. Applying them to arbitrary elements of $\mathcal{V}_1$ and $\mathcal{V}_2$ respectively and taking into account (2.74), we obtain

$$\begin{aligned}\langle A_1^* p_1(s), v\rangle_{\mathcal{V}_1',\mathcal{V}_1} &= \langle B_1 u_1(s) - B_2 u_2(s), B_1 v\rangle_{\mathcal{S}} \quad \forall v \in \mathcal{V}_1, \\ \langle A_2^* p_2(s), v\rangle_{\mathcal{V}_2',\mathcal{V}_2} &= -\langle B_1 u_2(s) - B_2 u_2(s), B_2 v\rangle_{\mathcal{S}} \quad \forall v \in \mathcal{V}_2.\end{aligned} \tag{2.76}$$

For the left-hand side we have

$$\langle A_1^* p_1(s), v\rangle_{\mathcal{V}_1',\mathcal{V}_1} = \langle p_1(s), A_1 v\rangle_{\mathcal{V}_1'',\mathcal{V}_1'} = \langle A_1 v, p_1(s)\rangle_{\mathcal{V}_1',\mathcal{V}_1} = \pi_1(v, p_1(s)) = \pi_1^*(p_1(s), v).$$

Similarly,

$$\langle A_2^* p_2(s), v\rangle_{\mathcal{V}_2',\mathcal{V}_2} = \pi_2^*(p_2(s), v).$$

Substituting these relations in (2.76), we discover that $p_1(s)$ and $p_2(s)$ satisfy (2.72) and (2.73), respectively.

The existence and uniqueness of solutions to (2.72) and (2.73) is established the same way as for Problems 2.5 and 2.6 (see Theorem 2.18 and Remark 2.9).

□

Thus, we obtain the following algorithm for calculating $\hat{J}'(s)$:

Algorithm 2.2 (Calculating $\hat{J}'(s)$).

1. For given s find u_1 and u_2 by solving Problems 2.5 and 2.6, i.e.

$$\begin{aligned} \pi_1(u_1, v) &= \ell v + sB_1 v \quad \forall v \in \mathcal{V}_1, \\ \pi_2(u_2, v) &= -sB_2 v \quad \forall v \in \mathcal{V}_2. \end{aligned}$$

2. Find p_1 and p_2 by solving the adjoint equations

$$\begin{aligned} \pi_1^*(p_1, v) &= \langle B_1 u_1 - B_2 u_2, B_1 v\rangle_S \quad \forall v \in \mathcal{V}_1, \\ \pi_2^*(p_2, v) &= -\langle B_1 u_1 - B_2 u_2, B_2 v\rangle_S \quad \forall v \in \mathcal{V}_2. \end{aligned}$$

3. Calculate

$$\hat{J}'(s) = B_1 p_1 - B_2 p_2.$$

Remark 2.21. The well-posedness of the discrete versions of the adjoint problems can be established the same way as it was done for the direct problems (see Section 2.6.1). The adjoint problems can also be solved in parallel, in the same finite element setting as the direct ones.

2.7 Simulation results

In this section we present the results of numerical simulations of the biosensor in one of the typical configurations. The x_1-length of the sensor is 1.8 mm; x_2-width is 2 mm. The thicknesses of the layers are given in Table 2.1.

The material properties are presented in Table 2.2. The tensors $\hat{G}, \hat{P}, G$ and e are specified in the Voigt notation. The material constants of the piezoelectric substrate are given in the crystallographic coordinate system. In order to get a surface shear wave travelling in x_1-direction, we take a so-called ST-cut of the quartz crystal. This cut is made by rotating the crystal to the angle of 42.75° around x_1-axis. The effective material

Layer	Thickness
Aptamer layer	0.004 μm
Shielding layer	0.1 μm
Guiding layer	5.5 μm
Substrate	0.3 mm

Table 2.1: Thickness of the layers

parameters, i.e. tensors G, e, and ε, are computed from the ones specified in Table 2.2 by the corresponding coordinate transformation.

The waves are excited by the 7 pairs of the alternating input electrodes. As described in Section 2.3 one set of them is grounded, another one is supplied with the alternating current of the form

$$\varphi(\boldsymbol{x}, t) = V_0 \sin \omega t,$$

where ω is the circular frequency related to the ordinary frequency f by

$$\omega = 2\pi f.$$

We have done a series of simulations with the amplitude voltage V_0 of 20 Volt and the frequency f varying from 75 to 125 MHz. The excited wave travels in x_1-direction and induces a potential difference on the 7 pairs of the alternating output electrodes on the other end of the sensor. Though the electrodes in each subset are not connected geometrically in our simulation, the electric potential on them is kept the same by the linking the corresponding degrees of freedom.

The result of a simulation is a Ritz-Galerkin solution to Problem 2.4. Recall that it is a vector of the form

$$(\boldsymbol{v}^{(1)}, \boldsymbol{v}^{(2)}, \varphi^{(1)}, \varphi^{(2)}).$$

The time-dependent displacements in the solid layers and the electric potential in the substrate are then constructed by (see (2.25) and (2.27))

$$\begin{aligned} \boldsymbol{u}(\boldsymbol{x}, t) &= \boldsymbol{v}^{(1)}(\boldsymbol{x}) \sin \omega t + \boldsymbol{v}^{(2)}(\boldsymbol{x}) \cos \omega t \qquad \text{in } \Omega_{pgsa}, \\ \varphi(\boldsymbol{x}, t) &= \varphi^{(1)}(\boldsymbol{x}) \sin \omega t + \varphi^{(2)}(\boldsymbol{x}) \cos \omega t \qquad \text{in } \Omega_p. \end{aligned}$$

Figure 2.5 shows the result of the simulation at the frequency 100 MHz. All the layers above the substrate are made invisible. The damping area is designated by the blue line.

Layer	Material	Density [kg/m³]	Other parameters
Liquid	Water	1000	Dynamic viscosity: 0.001 Pa·s Compressibilty: $4.6 \cdot 10^{-10}$ m²/N
Aptamer	Homogenized	1720	Tensor $\hat{G}$ [10^9 N/m²] : $\begin{array}{rrrrrr} -0.7193717 & 3.357842 & 2.066848 & 0 & 0 & 0 \\ 3.357842 & -0.7193717 & 2.066391 & 0 & 0 & 0 \\ 2.066848 & 2.066391 & 3.047297 & 0 & 0 & 0 \\ 0 & 0 & 0 & 0 & 0 & 0 \\ 0 & 0 & 0 & 0 & 0 & 0 \\ 0 & 0 & 0 & 0 & 0 & 0 \end{array}$ Tensor $\hat{P}$ [10^{-2} N · s/m²] : $\begin{array}{rrrrrr} 10.304 & 4.9644 & 6.0015 & 0 & 0 & 0 \\ 4.9644 & 12.065 & 5.6766 & 0 & 0 & 0 \\ 6.0015 & 5.6766 & 15.893 & 0 & 0 & 0 \\ 0 & 0 & 0 & 3.6534 & 0 & 0 \\ 0 & 0 & 0 & 0 & 3.6533 & 0 \\ 0 & 0 & 0 & 0 & 0 & 2.7075 \end{array}$
Shielding	Gold	19300	Young's modulus 78 GPa, Poisson's ratio 0.44
Guiding	SiO_2	2200	Young's modulus 72 GPa, Poisson's ratio 0.17
Substrate	α-quartz	2650	Stiffness tensor G [10^9 N/m²] : $\begin{array}{rrrrrr} 86.74 & 6.99 & 11.91 & -17.91 & 0 & 0 \\ 6.99 & 86.74 & 11.91 & 17.91 & 0 & 0 \\ 11.91 & 11.91 & 107.2 & 0 & 0 & 0 \\ -17.91 & 17.91 & 0 & 57.94 & 0 & 0 \\ 0 & 0 & 0 & 0 & 57.94 & -17.91 \\ 0 & 0 & 0 & 0 & -17.91 & 39.875 \end{array}$ Piezoelectric tensor e [C/m²] : $\begin{array}{rrrrrr} 0.171 & -0.171 & 0 & -0.0407 & 0 & 0 \\ 0 & 0 & 0 & 0 & 0.0407 & -0.171 \\ 0 & 0 & 0 & 0 & 0 & 0 \end{array}$ Dielectric tensor ϵ [$10^{-12} F/m$]: $\begin{array}{rrr} 39.97 & 0 & 0 \\ 0 & 39.97 & 0 \\ 0 & 0 & 41.03 \end{array}$

Table 2.2: Material parameters of the layers

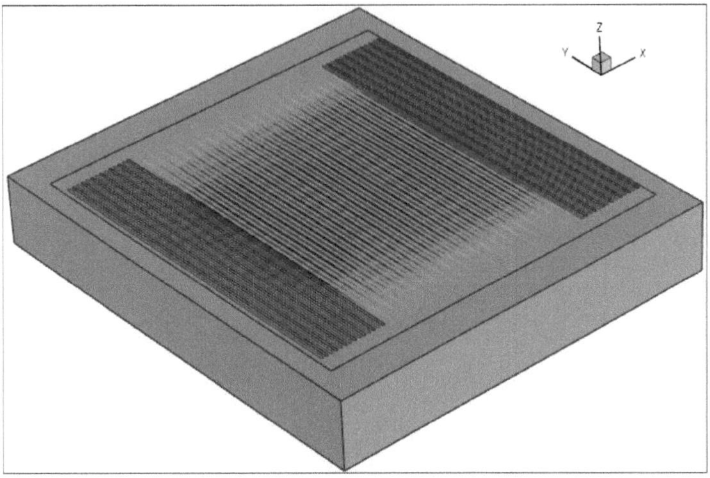

Figure 2.5: Shear component of $\boldsymbol{v}^{(1)}$ in the substrate.

The color scale represents the x_2 component of $\boldsymbol{v}^{(1)}$, i.e. the shear component of the wave. As expected it is almost periodic in x_1-direction and decays when approaching the damping area. The shear component of $\boldsymbol{v}^{(2)}$ has the same periodic structure as can be seen from Figure 2.6. This figure shows the shear components of both $\boldsymbol{v}^{(1)}$ and $\boldsymbol{v}^{(2)}$ evaluated on the middle line of the guiding layer parallel to the x_1-axis.

The decay rate of the wave when moving away downward from the substrate surface can be observed in Figure 2.7. It shows a cross-section of the substrate and the guiding layers by a vertical plane parallel to the x_1-axis. One can see that the amplitude of the displacements is the highest in the guiding layer and decays very quickly in the substrate. Hence the computed wave is a surface wave.

Figure 2.8 provides a vector view for $\boldsymbol{v}^{(1)}$ at the same cross-section. As expected the x_1 and x_3 components of $\boldsymbol{v}^{(1)}$ are negligible small in comparison with $v_2^{(1)}$. The same holds for $\boldsymbol{v}^{(2)}$. Thus the computed wave is a shear surface wave with displacements parallel to the x_2-axis.

Every pair of the input electrodes excites a wave with the wavelength depending on the frequency. Whenever the wavelength covers the distance between the pairs positions a whole number of times, the phases of the waves excited by the pairs coincide and resonance

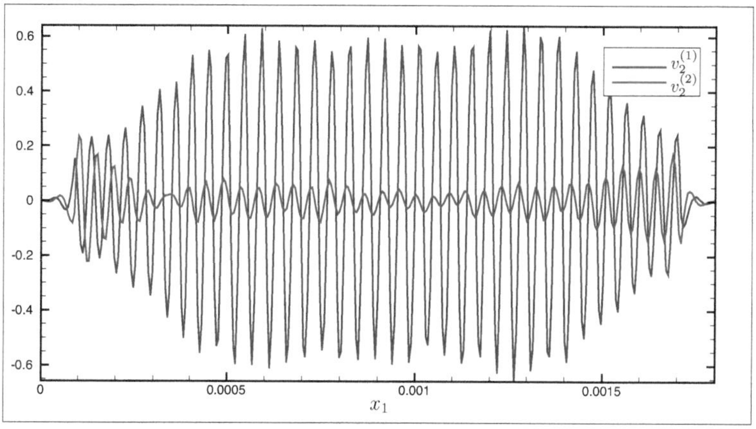

Figure 2.6: Shear components of $\boldsymbol{v}^{(1)}$ and $\boldsymbol{v}^{(2)}$ on the middle line in the guiding layer.

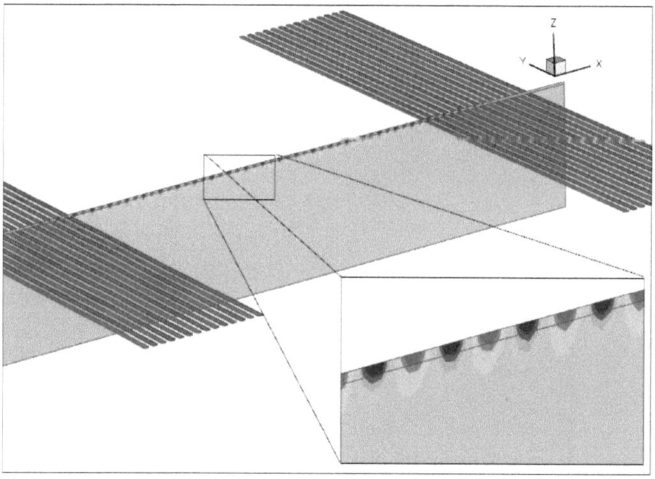

Figure 2.7: Component $v_2^{(1)}$ in the guiding layer and substrate evaluated at a cross section.

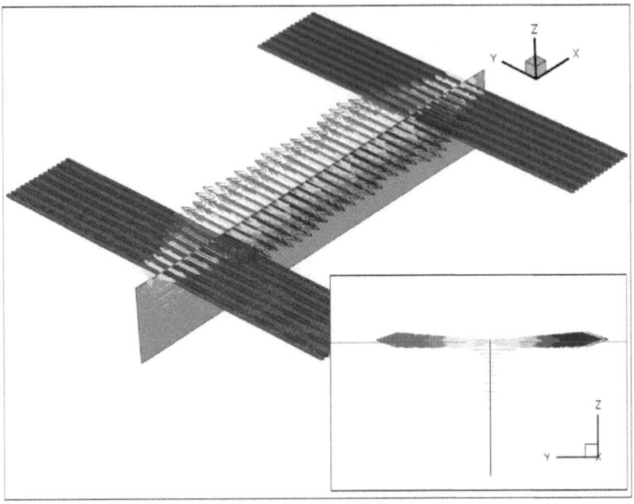

Figure 2.8: Vector $\boldsymbol{v}^{(1)}$ in the guiding layer and substrate evaluated at a cross section.

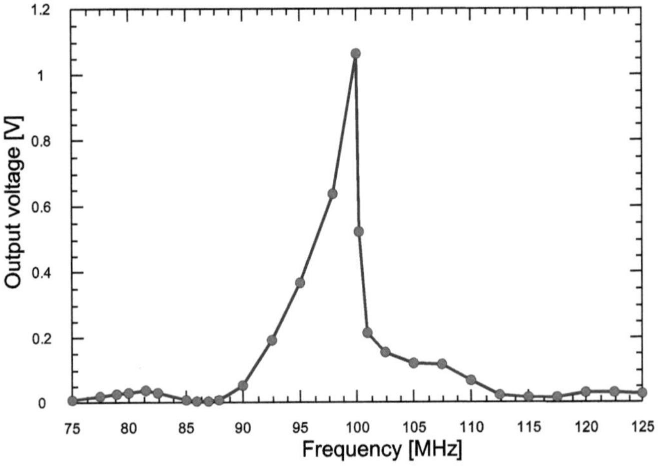

Figure 2.9: Output voltage.

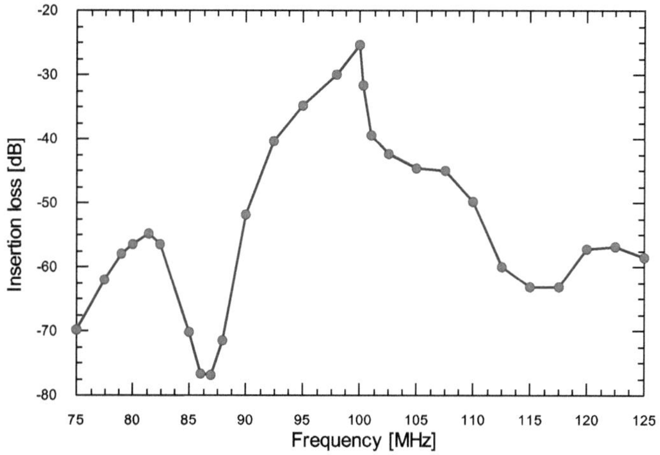

Figure 2.10: Insertion loss.

takes place. A frequency at which such a wavelength is achieved is a resonant frequency. This is also the operational frequency of the biosensor because the loss of the signal power is minimal in this case. To find out the resonant frequency we have performed a series of simulations with different values of the frequency and computed the so-called insertion loss that is defined as

$$20 \lg \frac{V}{V_0},$$

where V_0 is the voltage at the input electrodes as described above and V is the voltage at the output electrodes. In our case the output electrodes belonging to S_3 are grounded and hence the output voltage is computed by

$$V = \sqrt{\left(\varphi^{(1)}\big|_{S_4}\right)^2 + \left(\varphi^{(2)}\big|_{S_4}\right)^2}.$$

We set both the distance between the electrodes and the electrodes width to 11.25 μm. The difference in x_1 position between two neighboring pairs of electrodes is then 45 μm. The method described in Chapter 3 predicts that the wavelength of 45 μm is achieved at the frequency of 100 MHz (see Section 3.10.2 and Figure 3.15), i.e., the frequency of 100 MHz is resonant.

Our simulations by the finite element method confirm this. Figure 2.9 shows the ouput voltage for different values of the frequency. The insertion loss is depicted in Figure 2.10. Indeed, the highest value of the insetion loss is achieved at 100 Mhz.

The simulations have been carried out with a FE-Program called `FeliCs` that has been developed at TU München. The sparse solver `PARDISO` (see [61]) has been used for solving the arising linear systems. It proved to be quite efficient on shared memory multiprocessors giving a speedup up to ten times at 24 threads. The mesh size varied from $320 \times 50 \times 9$ to $440 \times 100 \times 20$ nodes yielding more than two millions of degrees of freedom.

3 Dispersion Relations in Multi-Layered Structures

3.1 Introduction

This chapter presents an approach to fast characterization of plain acoustic waves propagating in multi-layered medium contacting with a fluid or dielectric medium. It covers all the range of topics from the statement of the mathematical model up to the computer implementation. The underlying assumption of the approach is the infiniteness of layers in the horizontal directions.

Solid multi-layered structures considered here consist of any finite number of flat layers stacked together. The top and bottom layers are either semi-infinite or contact with a medium such as liquid, gas or vacuum. It is assumed throughout the work that the magnitudes considered are sufficiently small so that the linear laws describe the phenomena well enough.

The chapter is organized as follows. Section 3.3 presents the mathematical model and the algorithm for the calculation of plane acoustic waves in multi-layered elastic structures. The algorithm is first stated for a simple two-layers structure consisting of a semi-infinite half-space and a layer of finite thickness on top of it. The presented ideas are then generalized to the case of general multi-layered structures. Section 3.4 extends the approach to the case of piezoelectric layers. Sections 3.5 and 3.6 describe how the electro-mechanical influence of the contacting dielectric or liquid media is taken into account. Section 3.7 shows how the model from [28] for treating bristle-like solid-fluid interfaces can be adapted for use with the developed algorithm. Section 3.8 is devoted to accounting for so-called periodic multilayers. Section 3.9 describes the algorithm for building a dispersion curve on a range of frequencies starting with a single point found at given frequency. Finally, Section 3.10 is devoted to the computer implementation of the model. It consists of two parts. The first part discusses numerical issues such as computational complexity and avoiding numerical instabilities. The second part demonstrates the computer program implementing the model on an example of a real structure.

3.2 Notation

We use the cartesian coordinates (x_1, x_2, x_3). The direction of the wave propagation is taken as the x_1-axis. The x_3-axis is orthogonal to the layers surfaces; the x_2-axis is parallel to the surface and orthogonal to x_1 as shown in Figure 3.1. If not stated otherwise, the original of the coordinate system lies on the top surface of the bottom layer.

We number all the layers in the structure under consideration bottom-upwards starting with 1 corresponding to the bottom layer (which may occupy a half-space). The thickness of the n-th layer is denoted by $h^{(n)}$. The thickness of semi-infinite layers is $+\infty$. The x_3-coordinate of the top surface of the n-th layer is denoted by τ_n, i.e.

$$\tau_n := \sum_{j=2}^{n-1} h^{(j)}, \quad n \geqslant 1.$$

Note that τ_n is also the x_3-coordinate of the bottom surface of the $(n+1)$-th layer (if present). If the first layer is not semi-infinite, assume $\tau_0 := -h^{(1)}$.

The area occupied by the whole structure is denoted by $\Omega \subseteq \mathbb{R}^2$. It is unbounded in the x_1- and x_2-directions. The open area corresponding to the n-th layer is denoted by Ω_n. Obviously,

$$\overline{\Omega} = \bigcup_{n=1}^{N} \overline{\Omega}_n,$$

where N is the number of layers. We use the superscript (n) to specify that the corresponding variable belongs to the n-th layer. For example, $\boldsymbol{u}^{(n)}$ is the displacement vector in n-th layer, $\boldsymbol{u}^{(n)}$ is defined on $\overline{\Omega}_n$.

Vectors are distinguished from scalar quantities by writing them in a bold font.

The Einstein's summation convention is exploited throughout the work.

3.3 Elastic Multi-Layered Structures

3.3.1 Analysis of plane waves in elastic media

As mentioned above, we assume the displacements to be small so that the linear theory of elasticity is applicable. In particular, we make use of Hook's law that states that

$$\sigma_{ij} = G_{ijkl}\varepsilon_{kl}, \tag{3.1}$$

where σ and ε are the stress and the strain tensor respectively; G is the stiffness tensor. We assume the strain tensor ε to be infinitesimal, i.e.

$$\varepsilon(\boldsymbol{u}) = \frac{1}{2}(\nabla\boldsymbol{u} + (\nabla\boldsymbol{u})^T), \tag{3.2}$$

where $\boldsymbol{u}$ is the displacement vector. As follows from mechanical considerations, the stiffness tensor G possesses the following symmetry properties

$$G_{ijkl} = G_{klij} = G_{jikl}. \tag{3.3}$$

Besides, it is positive definite, i.e.

$$G_{ijkl}\xi_{ij}\xi_{kl} > 0, \tag{3.4}$$

for all symmetric non-zero tensors ξ.

Remark. The positiveness property (3.4) holds also for arbitrary non-antisymmetric tensors ξ. This can easily be shown using the symmetry property (3.3). Indeed,

$$G_{ijkl}\xi_{ij}\xi_{kl} = 2\,G_{ijkl}(\xi_{ij} + \xi_{ji})\xi_{kl} = 4\,G_{ijkl}(\xi_{ij} + \xi_{ji})(\xi_{kl} + \xi_{lk}) > 0.$$

The momentum conservation law gives the following relation (no body force is assumed):

$$\varrho\boldsymbol{u}_{tt} - \mathrm{div}\sigma = 0, \tag{3.5}$$

where ϱ is the mass density, and u_{tt} is the second derivative of the displacement vector $\boldsymbol{u}$ with respect to time t. Combining (3.5) with (3.1) and (3.2) and taking into account the symmetry properties of G, we obtain the following governing equation:

$$\varrho \boldsymbol{u}_{tt} - \operatorname{div}(G\nabla \boldsymbol{u}) = 0.$$

Rewritten in terms of components, it takes the form

$$\varrho \frac{\partial^2 u_i}{\partial t^2} - G_{ijkl} \frac{\partial^2 u_k}{\partial x_l \partial x_j} = 0, \quad i = 1, 2, 3. \tag{3.6}$$

A plane wave that travels in x_1-direction has a displacement field of the form

$$\boldsymbol{u}(x_1, x_3) = \Re\left\{\hat{\boldsymbol{c}}(x_3) e^{\mathrm{i}(\kappa x_1 - \omega t)}\right\},$$

where κ is the wave number and ω is the angular frequency. The phase velocity v is defined then as the ratio $\frac{\omega}{\kappa}$. Here i is the standard imaginary unit, $\Re\{\}$ denotes the real part of the expression in the parenthesis. The function $\hat{\boldsymbol{c}}$ is a complex-valued amplitude function. It is, however, more convenient to get rid of the complex part explicitly and to use the following form:

$$\boldsymbol{u}(x_1, x_3) = \boldsymbol{a}(x_3)\cos(\kappa x_1 - \omega t) + \boldsymbol{b}(x_3)\sin(\kappa x_1 - \omega t). \tag{3.7}$$

The amplitude vector functions $\boldsymbol{a}$ and $\boldsymbol{b}$ are real-valued now. Substituting (3.7) into (3.6) yields the following system of 6 ordinary differential equations:

$$\begin{cases} -G_{i3k3}\, \ddot{a}_k - (G_{i1k3} + G_{i3k1})\, \dot{b}_k + G_{i1k1}\, a_k - \varrho v^2\, a_i = 0, & i = 1, 2, 3, \\ -G_{i3k3}\, \ddot{b}_k + (G_{i1k3} + G_{i3k1})\, \dot{a}_k + G_{i1k1}\, b_k - \varrho v^2\, b_i = 0, & i = 1, 2, 3. \end{cases}$$

Here the dot denotes the differentiation with respect to the variable $\tilde{x}_3 = \kappa x_3$. The introduction of the variable $\tilde{x}_3$ is caused by numerical considerations discussed in details in Section 3.10.1. The system can be rewritten in matrix form as follows:

$$\begin{cases} -G_{\cdot 3 \cdot 3}\ddot{\boldsymbol{a}} - (G_{\cdot 1 \cdot 3} + G_{\cdot 3 \cdot 1})\, \dot{\boldsymbol{b}} + \left(G_{\cdot 1 \cdot 1} - \varrho v^2\, \mathbf{I}_3\right) \boldsymbol{a} = 0, \\ -G_{\cdot 3 \cdot 3}\ddot{\boldsymbol{b}} + (G_{\cdot 1 \cdot 3} + G_{\cdot 3 \cdot 1})\, \dot{\boldsymbol{a}} + \left(G_{\cdot 1 \cdot 1} - \varrho v^2\, \mathbf{I}_3\right) \boldsymbol{b} = 0. \end{cases} \tag{3.8}$$

Here the dots at subscripts of G denote that the corresponding indices vary. In this case two indices are always fixed and the other two vary, thus forming a matrix. $\mathbf{I}_n \in \mathbb{R}^{n \times n}$ is the unit $n \times n$ matrix.

Remark. The matrix $G_{\cdot 3\cdot 3}$ is positive definite. This follows from the positiveness of tensor G. Indeed, for all $\boldsymbol{x} \in \mathbb{R}^3$ such that $\boldsymbol{x} \neq 0$, we have

$$\boldsymbol{x}^T G_{\cdot 3\cdot 3}\, \boldsymbol{x} = x_i\, G_{i3k3}\, x_k = G_{i3k3}\, x_i\, x_k = G_{ijkl}\, x_i \delta_{3j}\, x_k \delta_{3l},$$

where δ_{ij} is the Kronecker symbol. Introducing tensor $\xi_{ij} = x_i \delta_{3j}$ and noting that it is not anti-symmetric for non-zero $\boldsymbol{x}$, we obtain

$$\boldsymbol{x}^T G_{\cdot 3\cdot 3}\, \boldsymbol{x} = G_{ijkl}\, x_i \delta_{3j}\, x_k \delta_{3l} = G_{ijkl}\, \xi_{ij}\, \xi_{kl} > 0.$$

Thus $G_{\cdot 3\cdot 3}$ is positive definite and therefore not singular. The system (3.8) can then be rewritten as follows:

$$\begin{cases} \ddot{\boldsymbol{a}} = -G_{\cdot 3\cdot 3}^{-1} \left(G_{\cdot 1\cdot 3} + G_{\cdot 3\cdot 1}\right) \dot{\boldsymbol{b}} + G_{\cdot 3\cdot 3}^{-1} \left(G_{\cdot 1\cdot 1} - \varrho v^2\, \mathbf{I}_3\right) \boldsymbol{a}, \\ \ddot{\boldsymbol{b}} = \quad G_{\cdot 3\cdot 3}^{-1} \left(G_{\cdot 1\cdot 3} + G_{\cdot 3\cdot 1}\right) \dot{\boldsymbol{a}} + G_{\cdot 3\cdot 3}^{-1} \left(G_{\cdot 1\cdot 1} - \varrho v^2\, \mathbf{I}_3\right) \boldsymbol{b}. \end{cases} \tag{3.9}$$

With the state vectors

$$\boldsymbol{s} := \begin{pmatrix} \boldsymbol{a} \\ \boldsymbol{b} \\ \dot{\boldsymbol{a}} \\ \dot{\boldsymbol{b}} \end{pmatrix}$$

the above system can be rewritten in the normal form as follows:

$$\dot{\boldsymbol{s}} = A\, \boldsymbol{s}, \tag{3.10}$$

where $A \in \mathbb{R}^{12\times 12}$ has the following structure:

$$\left(\begin{array}{cc|cc} 0 & 0 & \mathbf{I}_3 & 0 \\ 0 & 0 & 0 & \mathbf{I}_3 \\ \hline G_{\cdot 3\cdot 3}^{-1} \left(G_{\cdot 1\cdot 1} - \varrho v^2\, \mathbf{I}_3\right) & 0 & 0 & -G_{\cdot 3\cdot 3}^{-1} \left(G_{\cdot 1\cdot 3} + G_{\cdot 3\cdot 1}\right) \\ 0 & G_{\cdot 3\cdot 3}^{-1} \left(G_{\cdot 1\cdot 1} - \varrho v^2\, \mathbf{I}_3\right) & G_{\cdot 3\cdot 3}^{-1} \left(G_{\cdot 1\cdot 3} + G_{\cdot 3\cdot 1}\right) & 0 \end{array}\right) \tag{3.11}$$

We establish now an important property of this system.

Proposition 3.1. *Let λ be an eigenvalue of the matrix A with the eigenvector $\boldsymbol{p} \in \mathbb{R}^{12}$.*

Let $\boldsymbol{p}_1, \boldsymbol{p}_2, \boldsymbol{p}_3, \boldsymbol{p}_4 \in \mathbb{R}^3$ be the three-dimensional components of $\boldsymbol{p}$ such that

$$\boldsymbol{p} = \begin{pmatrix} \boldsymbol{p}_1 \\ \boldsymbol{p}_2 \\ \boldsymbol{p}_3 \\ \boldsymbol{p}_4 \end{pmatrix}.$$

Then $(-\lambda)$ is also an eigenvalue of A with the eigenvector $\hat{\boldsymbol{p}}$ in the form:

$$\hat{\boldsymbol{p}} = \begin{pmatrix} \boldsymbol{p}_2 \\ \boldsymbol{p}_1 \\ -\boldsymbol{p}_4 \\ -\boldsymbol{p}_3 \end{pmatrix}.$$

Proof. The statement can be verified straightforwardly. Nevertheless, we provide here a more constructive proof based on symmetry considerations.

From the physical point of view it is clear that if there is a solution of (3.10) decreasing with depth, i.e. as $x_3 \to -\infty$, then there must be a paired solution decreasing as $x_3 \to +\infty$, since the properties of the material are the same in the both directions.

To realize this idea mathematically, we define functions

$$\hat{\boldsymbol{a}}(x_3) := \boldsymbol{b}(-x_3),$$
$$\hat{\boldsymbol{b}}(x_3) := \boldsymbol{a}(-x_3)$$

and rewrite the system (3.9) for these new functions. It can easily be seen that the system for $\hat{\boldsymbol{a}}$ and $\hat{\boldsymbol{b}}$ has the same coefficients as the original one. This implies that for the state vector $\hat{\boldsymbol{s}}$ defined as

$$\hat{\boldsymbol{s}} := \begin{pmatrix} \hat{\boldsymbol{a}} \\ \hat{\boldsymbol{b}} \\ \dot{\hat{\boldsymbol{a}}} \\ \dot{\hat{\boldsymbol{b}}} \end{pmatrix}$$

the normal form is written with the same matrix A, i.e.

$$\dot{\hat{\boldsymbol{s}}} = A\,\hat{\boldsymbol{s}}.$$

Therefore, if λ is an eigenvalue of A with the eigenvector $\boldsymbol{p}$, then the system has a solution

$$\begin{pmatrix} \hat{\boldsymbol{a}}(x_3) \\ \hat{\boldsymbol{b}}(x_3) \\ \dot{\hat{\boldsymbol{a}}}(x_3) \\ \dot{\hat{\boldsymbol{b}}}(x_3) \end{pmatrix} = \begin{pmatrix} \boldsymbol{p_1} \\ \boldsymbol{p_2} \\ \boldsymbol{p_3} \\ \boldsymbol{p_4} \end{pmatrix} e^{\lambda\kappa x_3}.$$

Hence,

$$\boldsymbol{s}(x_3) = \begin{pmatrix} \boldsymbol{a}(x_3) \\ \boldsymbol{b}(x_3) \\ \dot{\boldsymbol{a}}(x_3) \\ \dot{\boldsymbol{b}}(x_3) \end{pmatrix} = \begin{pmatrix} \hat{\boldsymbol{b}}(-x_3) \\ \hat{\boldsymbol{a}}(-x_3) \\ -\dot{\hat{\boldsymbol{b}}}(-x_3) \\ -\dot{\hat{\boldsymbol{a}}}(-x_3) \end{pmatrix} = \begin{pmatrix} \boldsymbol{p_2} \\ \boldsymbol{p_1} \\ -\boldsymbol{p_4} \\ -\boldsymbol{p_3} \end{pmatrix} e^{\lambda\kappa(-x_3)} = \hat{\boldsymbol{p}} e^{-\lambda\kappa x_3}.$$

This is possible only if $(-\lambda)$ is an eigenvalue of A with the eigenvector $\hat{\boldsymbol{p}}$.

□

The general solution of the system (3.10) is of the form

$$\boldsymbol{s}(x_3) = \sum_{j=1}^{12} \tilde{c}_j \boldsymbol{p}_c^j e^{\lambda_j \kappa x_3},$$

where $\{\lambda_j\}_{j=1}^{12} \subset \mathbb{C}$ and $\{\boldsymbol{p}_c^j\}_{j=1}^{12} \subset \mathbb{C}^{12}$ are eigenvalues and eigenvectors of A, respectively; $\{\tilde{c}_j\}_{j=1}^{12}$ are arbitrary constants.

In order to get rid of complex solutions, every combination of complex conjugated eigenvalues

$$\tilde{c}_1 \left(\boldsymbol{p}_c^1 + i\boldsymbol{p}_c^2\right) e^{(\lambda_1 + \mathrm{i}\lambda_2)\kappa x_3} + \tilde{c}_2 \left(\boldsymbol{p}_c^1 - i\boldsymbol{p}_c^2\right) e^{(\lambda_1 - \mathrm{i}\lambda_2)\kappa x_3}$$

is replaced with the equivalent combination

$$c_1 \left[\boldsymbol{p}_c^1 \cos(\lambda_1 \kappa x_3) - \boldsymbol{p}_c^2 \sin(\lambda_2 \kappa x_3)\right] + c_2 \left[\boldsymbol{p}_c^2 \cos(\lambda_1 \kappa x_3) + \boldsymbol{p}_c^1 \sin(\lambda_2 \kappa x_3)\right],$$

where $c_1 = \tilde{c}_1 + \tilde{c}_2, c_2 = \mathrm{i}(\tilde{c}_1 - \tilde{c}_2)$. After renumbering the eigenvalues and eigenvectors in such a way that the real ones go first and the complex-conjugated follow one another, the general solution takes the following form:

$$\begin{aligned} \boldsymbol{s}(x_3) = & \sum_{j=1}^{n} c_j\, \boldsymbol{p}_c^j e^{\lambda_j \kappa x_3} + \\ & + \sum_{j=\frac{n}{2}+1}^{6} \Big(c_{2j-1} \Big[\Re \boldsymbol{p}_c^{2j} \cos(\Re\lambda_{2j}\kappa x_3) - \Im \boldsymbol{p}_c^{2j} \sin(\Im\lambda_{2j}\kappa x_3) \Big] + \\ & \qquad + c_{2j} \Big[\Im \boldsymbol{p}_c^{2j} \cos(\Re\lambda_{2j}\kappa x_3) + \Re \boldsymbol{p}_c^{2j} \sin(\Im\lambda_{2j}\kappa x_3) \Big] \Big), \end{aligned} \tag{3.12}$$

where $c_j = \tilde{c}_j$ for the real eigenvalues; n is the number of the real eigenvalues; $\Im$ denotes taking the imaginary part. Note that n is always even. The components in the first sum are exponential solutions, the ones in the second sum are oscillating solutions. In order to make the notation less bulky, we denote the multipliers of c_j by $\boldsymbol{p}^j(x_3)$, i.e.

$$\boldsymbol{p}^j(x_3) = \begin{cases} \boldsymbol{p}_c^j e^{\lambda_j \kappa x_3} & \text{if } 1 \leqslant j \leqslant n, \\ \Re \boldsymbol{p}_c^j \cos(\Re\lambda_j \kappa x_3) - \Im \boldsymbol{p}_c^j \sin(\Im\lambda_j \kappa x_3) & \text{if } j > n \text{ and } j \text{ is even,} \\ \Re \boldsymbol{p}_c^{j+1} \cos(\Re\lambda_{j+1} \kappa x_3) - \Im \boldsymbol{p}_c^{j+1} \sin(\Im\lambda_{j+1} \kappa x_3) & \text{if } j > n \text{ and } j \text{ is odd.} \end{cases} \tag{3.13}$$

The general solution is then of the form

$$\boldsymbol{s}(x_3) = \sum_{j=1}^{12} c_j \boldsymbol{p}^j(x_3). \tag{3.14}$$

Extracting the components of the state vector $\boldsymbol{s}$ corresponding to $\boldsymbol{a}$ and $\boldsymbol{b}$, and substituting them into (3.7), we obtain

$$\boldsymbol{u}(x_1, x_3) = \sum_{j=1}^{12} c_j \boldsymbol{f}^j(x_3) \cdot \cos(\kappa x_1 - \omega t) + \sum_{j=1}^{12} c_j \boldsymbol{g}^j(x_3) \cdot \sin(\kappa x_1 - \omega t), \tag{3.15}$$

where the real-valued vector functions $\boldsymbol{f}^j(x_3), \boldsymbol{g}^j(x_3)$ contain the components of $\boldsymbol{p}^j(x_3)$ corresponding to $\boldsymbol{a}$ and $\boldsymbol{b}$, respectively.

3.3.2 Two-layered structure

The algorithm lying in the base of the model is demonstrated here on an example of a structure consisting of a semi-infinite half-space solid (further refereed as substrate) coated with an overlying plate (see Figure 3.1). Only mechanical elastic properties of materials are taken into account here (piezoelectric materials are considered in Section 3.4). No contact

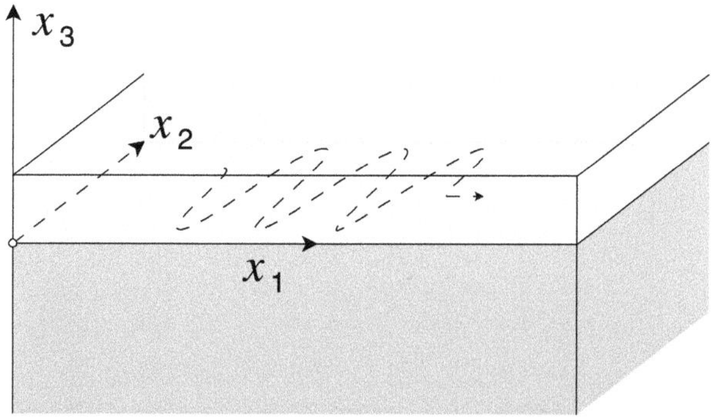

Figure 3.1: A sample structure. Half-space solid coated with an elastic layer.

medium is assumed.

Applying the expression (3.15) to our two layers, we get

$$\begin{cases} \boldsymbol{u}^{(1)}(x_1,x_3) = \sum_{j=1}^{12} c_j^{(1)} \boldsymbol{f}^{(1),j}(x_3)\cdot\cos(\kappa x_1 - \omega t) + \sum_{j=1}^{12} c_j^{(1)} \boldsymbol{g}^{(1),j}(x_3)\cdot\sin(\kappa x_1 - \omega t), \\ \boldsymbol{u}^{(2)}(x_1,x_3) = \sum_{j=1}^{12} c_j^{(2)} \boldsymbol{f}^{(2),j}(x_3)\cdot\cos(\kappa x_1 - \omega t) + \sum_{j=1}^{12} c_j^{(2)} \boldsymbol{g}^{(2),j}(x_3)\cdot\sin(\kappa x_1 - \omega t). \end{cases} \tag{3.16}$$

The solutions $\boldsymbol{u}^{(1)}$ and $\boldsymbol{u}^{(2)}$ in the substrate and the top layer respectively are coupled by the conditions on the interface between them. Together with the boundary conditions they determine the admissible values for constants $\{c_j^{(1)}\}_{j=1}^{12}$ and $\{c_j^{(2)}\}_{j=1}^{12}$, thus filtering out the wave solutions infeasible in the whole structure.

Interface conditions. The interface conditions include the continuity of the displacement field and the pressure equilibrium at the plane $x_3 = \tau_1 (:= 0)$, i.e.

$$\boldsymbol{u}^{(1)} = \boldsymbol{u}^{(2)},$$
$$\sigma^{(1)}\boldsymbol{n} = \sigma^{(2)}\boldsymbol{n},$$

where

$$\boldsymbol{n} := \begin{pmatrix} 0 \\ 0 \\ 1 \end{pmatrix}$$

is the unit vector normal to the interface plane.

Boundary conditions. As assumed the structure has no contacting media, which means that there are no external forces acting on the top surface of it. Hence the only boundary condition at the plane $x_3 = h^{(2)}$ is the absence of the pressure, i.e.

$$\sigma^{(2)}\boldsymbol{n} = 0.$$

In the substrate we require that the amplitude of the wave decays with the depth. Mathematically it means that $\boldsymbol{a}^{(1)}(x_3), \boldsymbol{b}^{(1)}(x_3) \to 0$ as $x_3 \to -\infty$. This can only be achieved if no terms with $\Re\lambda_i^{(1)} \leqslant 0$ are present in the first equations of (3.16). That means that

$$c_i^{(1)} = 0 \quad \text{for all } i \text{ such that } \Re\lambda_i^{(1)} \leqslant 0. \tag{3.17}$$

Notice that due to Proposition 3.1 this condition filters out at least a half of components in the sum.

Combining the interface and the boundary conditions, we obtain the following system of equations:

$$\begin{cases} \sigma^{(2)}\boldsymbol{n} = 0 & \text{at } x_3 = \tau_2, \\ \\ \boldsymbol{u}^{(1)} = \boldsymbol{u}^{(2)} & \text{at } x_3 = \tau_1, \\ \sigma^{(1)}\boldsymbol{n} = \sigma^{(2)}\boldsymbol{n} & \text{at } x_3 = \tau_1, \\ \\ c_j^{(1)} = 0 & \text{if } \Re\lambda_j^{(1)} \leqslant 0. \end{cases} \tag{3.18}$$

Using (3.1) and (3.2) this system can be rewritten in terms of components as follows:

$$\begin{cases} G_{i3kl}^{(2)} \dfrac{\partial u_k^{(2)}}{\partial x_l}\bigg|_{x_3=\tau_2} = 0, \quad i = 1,2,3, \\ \\ \left(u_i^{(1)} - u_i^{(2)}\right)\Big|_{x_3=\tau_1} = 0, \quad i = 1,2,3, \\ \left(G_{i3kl}^{(1)} \dfrac{\partial u_k^{(1)}}{\partial x_l} - G_{i3kl}^{(2)} \dfrac{\partial u_k^{(2)}}{\partial x_l}\right)\bigg|_{x_3=\tau_1} = 0, \quad i = 1,2,3, \\ \\ c_j^{(1)} = 0 \quad \text{if } \Re\lambda_j^{(1)} \leqslant 0. \end{cases} \tag{3.19}$$

Substituting $\boldsymbol{u}^{(1)}$ and $\boldsymbol{u}^{(2)}$ from (3.16) into the first three conditions and equating the coefficients of $\sin(\kappa x_1 - \omega t)$ and $\cos(\kappa x_1 - \omega t)$, we obtain a homogeneous system of 18 linear equations for unknown coefficients $\{c_i^{(1)}\}_{i=1}^{12}$ and $\{c_i^{(2)}\}_{i=1}^{12}$. The last condition in (3.19) makes sure that no more than 18 of them are present so that the system is not underdetermined in general. We denote the matrix of this system by F and refer to it as *the fitting matrix* of the structure. The name should suggest that it describes relations "fitting" wave solutions in single layers to each other. Further, denote by $\boldsymbol{c}$ the vector of all non-zero coefficients collected from the both layers. Then the system above takes the form

$$F\boldsymbol{c} = 0. \tag{3.20}$$

Note that in general F is not a square matrix. The system admits a non-trivial solution iff the following equivalent conditions are fulfilled:

- The matrix F is rank deficient.
- $\det(F^T F) = 0$.
- 0 is an eigenvalue of $F^T F$
- $\dfrac{\lambda_{min}(F^T F)}{\lambda_{max}(F^T F)}$=0, where $\lambda_{min}(F^T F)$ and $\lambda_{max}(F^T F)$ are minimal and maximal eigenvalues of $F^T F$, respectively.

From the computational point of view these conditions are equivalent, i.e. the checking of them is of the same computational complexity. We make a choice in favor of the last one, because it provides us with a quantity that shows the degree of singularity of $F^T F$, and in contrast to other quantities it does not scale when material parameters are scaled.

The matrix F depends on two variables, on the frequency ω and on the wave number κ. They related to each other by the phase velocity v as follows:

$$\kappa = \frac{\omega}{v}.$$

We assume now that the frequency ω is fixed. Then we can consider the phase velocity v as the independent variable. We investigate now the behavior of the system (3.20) as v varies. Let us define the function $f : \mathbb{R}^+ \to \mathbb{R}_0^+$ by the rule

$$f : v \mapsto \frac{\lambda_{min}(F^T(v)F(v))}{\lambda_{max}(F^T(v)F(v))}. \tag{3.21}$$

We will refer to this function as *the fitting function* of the structure. The structure admits a plane wave propagating in the x_1 direction with the phase velocity v iff

$$f(v) = 0.$$

Note that the matrix $F^T F$ is positive semidefinite. Hence its eigenvalues are non-negative and $f \geqslant 0$. Thus, the problem of finding feasible wave solutions is reduced to the problem of finding roots of a non-negative function. In practice, however, the calculation error prevents the fitting function from turning into zero. The calculated values usually remain positive but may approach zero very closely. Therefore, instead of seeking for roots of the function, it is more promising to look for its local minima close to zero. In general, the fitting function possesses no convexity properties and may have several minima corresponding to different wave modes.

If v is a root of the fitting function, the system (3.20) possesses a non-trivial solution that can be found as follows:

Proposition 3.2. *Let v be a root of the fitting function and $\mathbf{c}$ be the eigenvector of the matrix $F^T(v)F(v)$ corresponding to $\lambda_{min}(v)$. Then $\mathbf{c}$ solves the system (3.20).*

Proof. Since v is a root of the fitting function,

$$\lambda_{min}(v) = 0.$$

Then we have

$$\begin{aligned} & F^T(v)F(v)\mathbf{c} = \lambda_{min}(v)\mathbf{c} = 0 \\ \Rightarrow \quad & \mathbf{c}^T F^T(v)F(v)\mathbf{c} = 0 \\ \Rightarrow \quad & [F(v)\mathbf{c}]^T[F(v)\mathbf{c}] = 0 \\ \Rightarrow \quad & F(v)\mathbf{c} = 0. \end{aligned}$$

That is, $\mathbf{c}$ solves (3.20). □

When the vector of coefficients $\mathbf{c}$ is found, the displacement vectors $\boldsymbol{u}^{(1)}$ and $\boldsymbol{u}^{(2)}$ are determined by (3.16). The wave solution is thus completely constructed. Summing up all the steps we describe now the algorithm for finding plane wave solutions on a range of velocities.

Algorithm 3.1.

1. Choose frequency ω, range of velocities $[v_0, v_m]$ and some discretization

$$v_0 < v_1 < \cdots < v_m.$$

2. For each $i = 0, \ldots, m$ calculate $f(v_i)$ as follows:
   - For each layer $n = 1, 2$ do
     - Build the matrix $A^{(n)}(v_i)$ as defined in (3.11).
     - Find the eigenvalues and eigenvectors of $A^{(n)}(v_i)$.
     - If n-th layer is the semi-infinite, remove the components corresponding to non-decaying solutions (see (3.17)).
     - For complex eigenvalues extract the corresponding real solutions to obtain the representation (3.12) with unknown coefficients $c_j^{(n)}$ for the state vector $\boldsymbol{s}^{(n)}$.
     - On the base of $\boldsymbol{s}^{(n)}$ build the form (3.15) for $\boldsymbol{u}^{(n)}$.
       Build a similar form for $\nabla \boldsymbol{u}^{(n)}$ by differentiating (3.15).
   - On the base of boundary and interface conditions (3.19) assembly the fitting matrix $F(v_i)$.
   - Calculate $f(v_i)$ as defined in (3.21).
3. Pick up an interval of velocities U containing just one root of f.
4. Find the minimum of f in U.
5. For $v^* \in \arg\min_{v \in U} f(v)$ calculate the eigenvector $\mathbf{c}(v^*)$ corresponding to $\lambda_{min}(v^*)$.
6. Construct the wave solution by substituting $\mathbf{c}(v^*)$ in (3.16).

The algorithm is controlled by the user at two stages. First, the user chooses the preliminary range of velocities to calculate the fitting function on it. Usually the approximative range of feasible velocities is known. Second choice is the choice of a root if more than one is present on the calculated interval. Figure 3.2 shows an example of a fitting function with three roots corresponding to different wave types feasible in the structure.

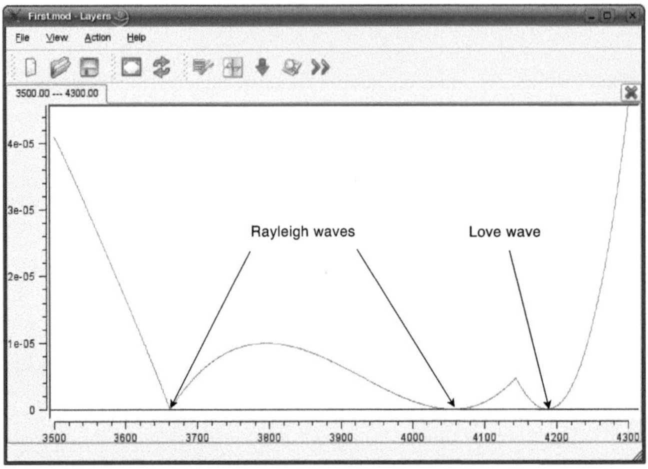

Figure 3.2: An example of fitting function. Three roots are the wave velocities for different wave modes feasible in the structure.

In further sections we adapt this algorithm to structures consisting of arbitrary number of layers of different types. When doing this, special attention is given to the checking that the number of unknown coefficients determining the wave in the whole structure is not bigger that the number of boundary and interface conditions. So that the fitting matrix does not become undetermined. Otherwise, a non-trivial solution to (3.20) would exist at any phase velocity and the algorithm would make no sense.

3.3.3 N-layered structure

Let us consider now a structure consisting of N stacked layers (see Figure 3.3). Like in the case of two-layered structure we obtain the representation (3.15) for each layer and derive the system (3.20) for the vector of unknown coefficients $\boldsymbol{c}$. Algorithm 3.1 can be used without significant changes. The only difference is the number of interface conditions and consequently the order of the matrix F. The set of the conditions is slightly different for three possible variants denoted in Figure 3.3 by (a), (b), and (c).

In the variant (a), the bottom layer is semi-infinite and the top layer is finite. The

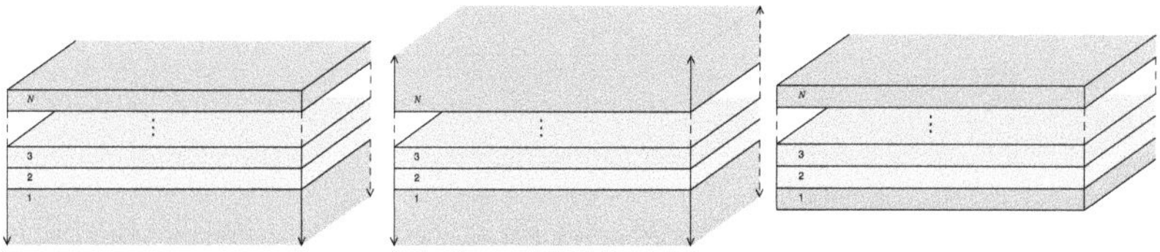

(a) Only the bottom layer is semi-infinite (b) The top and the bottom layers are semi-infinite (c) No semi-infinite layers

Figure 3.3: N-layered structures

conditions are as follows:

$$\begin{cases} \sigma^{(N)}\boldsymbol{n} = 0 & \text{at } x_3 = \tau_N, \\ \boldsymbol{u}^{(m-1)} = \boldsymbol{u}^{(m)} & \text{at } x_3 = \tau_{m-1}, \quad m = 2\dots N, \\ \sigma^{(m-1)}\boldsymbol{n} = \sigma^{(m)}\boldsymbol{n} & \text{at } x_3 = \tau_{m-1}, \quad m = 2\dots N, \\ c_j^{(1)} = 0 & \text{if } \Re\lambda_j^{(1)} \leqslant 0. \end{cases} \tag{3.22}$$

This system is the N-layered analog of the system (3.18). The first three lines yield $(1 + (N-1) \times 2) \times 3 \times 2 = 12N - 6$ scalar equations. The last factor 2 appears because the coefficients of $\sin(\kappa x_1 - \omega t)$ and $\cos(\kappa x_1 - \omega t)$ are equated. The number of unknown coefficients is 12 for each layer and $12N$ for the whole structure. Proposition 3.1 ensures that the last condition in (3.22) filters out at least a half of coefficients for the substrate. Hence the number of unknown coefficients does not exceed $12N - 6$.

Let us now consider the variant (b). The top and the bottom layers are now both semi-infinite (see Figure 3.3(b)). In this case, the condition on the top surface is replaced with the requirement that the amplitude decays as $x_3 \to +\infty$. Hence we have

$$\begin{cases} c_j^{(N)} = 0 & \text{if } \Re\lambda_j^{(N)} \geqslant 0, \\ \boldsymbol{u}^{(m-1)} = \boldsymbol{u}^{(m)} & \text{at } x_3 = \tau_{m-1}, \quad m = 2\dots N, \\ \sigma^{(m-1)}\boldsymbol{n} = \sigma^{(m)}\boldsymbol{n} & \text{at } x_3 = \tau_{m-1}, \quad m = 2\dots N, \\ c_j^{(1)} = 0 & \text{if } \Re\lambda_j^{(1)} \leqslant 0. \end{cases} \tag{3.23}$$

In comparison to the variant (a), the number of equations becomes 6 less. On the other hand, the condition of the amplitude decaying removes at least 6 unknown coefficients due to Proposition 3.1.

In the variant (c) no semi-infinite layers are present. Hence there are no amplitude decaying conditions. Instead, we have boundary conditions at the top and bottom surfaces. The system is then as follows:

$$\begin{cases} \sigma^{(N)}\boldsymbol{n} = 0 & \text{at } x_3 = \tau_N, \\ \boldsymbol{u}^{(m-1)} = \boldsymbol{u}^{(m)} & \text{at } x_3 = \tau_{m-1}, \quad m = 2\ldots N, \\ \sigma^{(m-1)}\boldsymbol{n} = \sigma^{(m)}\boldsymbol{n} & \text{at } x_3 = \tau_{m-1}, \quad m = 2\ldots N, \\ \sigma^{(1)}\boldsymbol{n} = 0 & \text{at } x_3 = \tau_0. \end{cases} \tag{3.24}$$

The number of equations in this case is equal to the number of the unknown coefficients $(= 12N)$.

Thus, in all the variants the number of unknowns does not exceed the number of equations. This means that the number of rows in the matrix F is not less than the number of columns and the system (3.20) is in general not underdetermined. This enables us to treat the multi-layered case by using the same approach based on the introduction of the fitting function.

3.4 Introducing Piezoelectricity

Since acoustic waves in sensors are usually excited by means of piezoelectric materials, the modeling of piezoelectric layers is extremely important. This section briefly describes the basic relations of the linear theory of piezoelectricity and shows how the algorithm described in the previous section is adjusted to deal with piezoelectric layers.

3.4.1 Analysis of plane waves in piezoelectric media

Again, we assume the displacements to be sufficiently small to justify the use of the linear theory of piezoelectricity. In piezoelectric materials the mechanical stress σ is caused not only by the mechanical strain ε but also by the electric field $\boldsymbol{E}$ (this is so-called **converse piezoelectric effect**). We assume the following constitutive relation:

$$\sigma_{ij} = G_{ijkl}\varepsilon_{kl} - e_{kij}E_k, \tag{3.25}$$

where e is the third-order piezoelectric tensor responsible for the coupling between mechanical strain and electric field. The coupling effect works in both directions, i.e. mechanical deformations generate electrical polarization in the material (**direct piezoelectric effect**). The constitutive relation for the electric displacements $\boldsymbol{D}$ is

$$D_i = \epsilon_{ij} E_j + e_{ikl} \varepsilon_{kl}, \tag{3.26}$$

where ϵ is the dielectric permittivity tensor.

It is easily seen from Maxwell's equations that the magnitude of the rotational component of the electric field is negligibly small. This follows from the smallness of the wave velocity in comparison with the speed of light. Therefore, we can represent the electric field as the gradient of a scalar potential function, i.e.

$$\boldsymbol{E} = -\nabla \phi. \tag{3.27}$$

The funciton ϕ is called the electric potential of the electric field $\boldsymbol{E}$. Furthermore, piezoelectric materials are insulators. Therefore no free volume charges may exist. This implies

$$\mathrm{div} \boldsymbol{D} = 0. \tag{3.28}$$

The tensors ϵ and e have the following symmetry properties:

$$\begin{aligned} \epsilon_{ij} &= \epsilon_{ji}, \\ e_{ikl} &= e_{ilk}, \quad i = 1, 2, 3. \end{aligned}$$

Besides, ϵ is positive-definite, i.e.

$$\forall \boldsymbol{\xi} \in \mathbb{R}^3 \setminus \{0\} \quad \epsilon_{ij} \xi_i \xi_j > 0.$$

Substituting (3.25) into (3.5) and (3.26) into (3.28), we obtain the system of coupled governing equations for the displacements $\boldsymbol{u}$ and the electric potential ϕ:

$$\begin{cases} \varrho \dfrac{\partial^2 u_i}{\partial t^2} - G_{ijkl} \dfrac{\partial^2 u_k}{\partial x_l \partial x_j} - e_{kij} \dfrac{\partial^2 \phi}{\partial x_k \partial x_j} = 0, \quad i = 1, 2, 3, \\ -\epsilon_{ij} \dfrac{\partial^2 \phi}{\partial x_j \partial x_i} + e_{ikl} \dfrac{\partial^2 u_k}{\partial x_l \partial x_i} = 0. \end{cases} \tag{3.29}$$

As in the case of elastic layers we are looking for plane wave solutions to this system in the form

$$\begin{aligned}\boldsymbol{u}(x_1,x_3) =& \boldsymbol{a}(x_3)\cos(\kappa x_1-\omega t)+\boldsymbol{b}(x_3)\sin(\kappa x_1-\omega t),\\ \phi(x_1,x_3) =& \alpha(x_3)\cos(\kappa x_1-\omega t)+\beta(x_3)\sin(\kappa x_1-\omega t).\end{aligned} \tag{3.30}$$

The unknown amplitude functions $\boldsymbol{a}$, $\boldsymbol{b}$, α and β are assumed to be real-valued here. Substituting (3.30) into (3.29) yields the following linear system of 8 ordinary differential equations:

$$\begin{cases} -G_{i3k3}\,\ddot{a}_k-(G_{i1k3}+G_{i3k1})\,\dot{b}_k+G_{i1k1}\,a_k-\varrho v^2\,a_i- \\ \qquad\qquad -e_{3i3}\,\ddot{\alpha}-(e_{1i3}+e_{3i1})\,\dot{\beta}+e_{1i1}\,\alpha=0, \quad i=1,2,3, \\ -G_{i3k3}\,\ddot{b}_k+(G_{i1k3}+G_{i3k1})\,\dot{a}_k+G_{i1k1}\,b_k-\varrho v^2\,b_i- \\ \qquad\qquad -e_{3i3}\,\ddot{\beta}+(e_{1i3}+e_{3i1})\,\dot{\alpha}+e_{1i1}\,\beta=0, \quad i=1,2,3, \\ \\ -\epsilon_{33}\,\ddot{\alpha}-(\epsilon_{13}+\epsilon_{31})\,\dot{\beta}+\epsilon_{11}\,\alpha+ \\ \qquad\qquad +e_{3k3}\,\ddot{a}_k+(e_{1k3}+e_{3k1})\,\dot{b}_k-e_{1k1}\,a_k=0, \\ -\epsilon_{33}\,\ddot{\beta}+(\epsilon_{13}+\epsilon_{31})\,\dot{\alpha}+\epsilon_{11}\,\beta+ \\ \qquad\qquad +e_{3k3}\,\ddot{b}_k-(e_{1k3}+e_{3k1})\,\dot{a}_k-e_{1k1}\,b_k=0. \end{cases}$$

Note that the dot here denotes the differentiation with respect to the variable $\tilde{x}_3 := \kappa x_3$. The system can be rewritten in matrix form as follows:

$$\left(\begin{array}{cc|cc} G_{\cdot 3\cdot 3} & \boldsymbol{e}_{3\cdot 3} & 0 & 0 \\ -\boldsymbol{e}_{3\cdot 3}^T & \epsilon_{33} & 0 & 0 \\ \hline 0 & 0 & G_{\cdot 3\cdot 3} & \boldsymbol{e}_{3\cdot 3} \\ 0 & 0 & -\boldsymbol{e}_{3\cdot 3}^T & \epsilon_{33} \end{array}\right)\begin{pmatrix}\ddot{\boldsymbol{a}}\\ \ddot{\alpha}\\ \ddot{\boldsymbol{b}}\\ \ddot{\beta}\end{pmatrix} = \tag{3.31}$$

$$= \left(\begin{array}{cc|cc} 0 & 0 & -(G_{\cdot 1\cdot 3}+G_{\cdot 3\cdot 1}) & -(\boldsymbol{e}_{1\cdot 3}+\boldsymbol{e}_{3\cdot 1}) \\ 0 & 0 & (\boldsymbol{e}_{1\cdot 3}^T+\boldsymbol{e}_{3\cdot 1}^T) & -(\epsilon_{13}+\epsilon_{31}) \\ \hline (G_{\cdot 1\cdot 3}+G_{\cdot 3\cdot 1}) & (\boldsymbol{e}_{1\cdot 3}+\boldsymbol{e}_{3\cdot 1}) & 0 & 0 \\ -(\boldsymbol{e}_{1\cdot 3}^T+\boldsymbol{e}_{3\cdot 1}^T) & (\epsilon_{13}+\epsilon_{31}) & 0 & 0 \end{array}\right)\begin{pmatrix}\dot{\boldsymbol{a}}\\ \dot{\alpha}\\ \dot{\boldsymbol{b}}\\ \dot{\beta}\end{pmatrix}+$$

$$+\left(\begin{array}{cc|cc} (G_{\cdot 1\cdot 1}-\varrho v^2\,\mathbf{I}_3) & \boldsymbol{e}_{1\cdot 1} & 0 & 0 \\ -\boldsymbol{e}_{1\cdot 1}^T & \epsilon_{11} & 0 & 0 \\ \hline 0 & 0 & (G_{\cdot 1\cdot 1}-\varrho v^2\,\mathbf{I}_3) & \boldsymbol{e}_{1\cdot 1} \\ 0 & 0 & -\boldsymbol{e}_{1\cdot 1}^T & \epsilon_{11} \end{array}\right)\begin{pmatrix}\boldsymbol{a}\\ \alpha\\ \boldsymbol{b}\\ \beta\end{pmatrix}.$$

Proposition 3.3. *The matrix*

$$\begin{pmatrix} G_{\cdot 3 \cdot 3} & \boldsymbol{e}_{3 \cdot 3} \\ -\boldsymbol{e}_{3 \cdot 3}^T & \epsilon_{33} \end{pmatrix} \in \mathbb{R}^{4 \times 4}$$

is positive-definite.

Proof. First note that $\epsilon_{33} > 0$. This follows from the positiveness of ϵ. Indeed, for $\boldsymbol{\xi} = (0,\, 0,\, 1)^T$ we get

$$\epsilon_{33} = \boldsymbol{\xi}^T \epsilon\, \boldsymbol{\xi} > 0.$$

Further, for all $\boldsymbol{x} \in \mathbb{R}^3, y \in \mathbb{R}$ such that $\left(\boldsymbol{x}^T, y\right)^T \in \mathbb{R}^4 \setminus \{0\}$, we have

$$\left(\boldsymbol{x}^T,\, y\right) \begin{pmatrix} G_{\cdot 3 \cdot 3} & \boldsymbol{e}_{3 \cdot 3} \\ -\boldsymbol{e}_{3 \cdot 3}^T & \epsilon_{33} \end{pmatrix} \begin{pmatrix} \boldsymbol{x} \\ y \end{pmatrix} = \left(\boldsymbol{x}^T,\, y\right) \begin{pmatrix} G_{\cdot 3 \cdot 3}\boldsymbol{x} + \boldsymbol{e}_{3 \cdot 3}\, y \\ -\boldsymbol{e}_{3 \cdot 3}^T \boldsymbol{x} + \epsilon_{33}\, y \end{pmatrix} =$$

$$= \boldsymbol{x}^T G_{\cdot 3 \cdot 3}\boldsymbol{x} + \boldsymbol{x}^T \boldsymbol{e}_{3 \cdot 3}\, y - y\, \boldsymbol{e}_{3 \cdot 3}^T \boldsymbol{x} + y\, \epsilon_{33}\, y = \boldsymbol{x}^T G_{\cdot 3 \cdot 3}\boldsymbol{x} + \epsilon_{33}\, y^2 > 0.$$

The last inequality follows from the positiveness of $G_{\cdot 3 \cdot 3}$ and ϵ_{33}.

□

Thus the matrix on the left-hand side of (3.31) is positive-definite and therefore invertible. Then we can rewrite the system (3.31) in the normal form as follows:

$$\dot{\boldsymbol{s}}_p = A_p\, \boldsymbol{s}_p, \tag{3.32}$$

where $\boldsymbol{s}_p$ is the state vector defined by

$$\boldsymbol{s}_p := \left(a_1, a_2, a_3, \alpha, b_1, b_2, b_3, \beta, \dot{a}_1, \dot{a}_2, \dot{a}_3, \dot{\alpha}, \dot{b}_1, \dot{b}_2, \dot{b}_3, \dot{\beta}\right)^T \in \mathbb{R}^{16},$$

and $A_p \in \mathbb{R}^{16 \times 16}$ is the resulting matrix. This system is the piezoelectric analog of the system (3.10) that we derived for elastic materials. In contrast to the elastic case, the system has additional terms and contains two more variables due to taking into account electric and piezoelectric effects. However, the general principle of treating the system remains the same. In particular, the equivalent of the Proposition (3.1) takes place.

Proposition 3.4. *Let λ be an eigenvalue of the matrix A_p with the eigenvector $\boldsymbol{p} \in \mathbb{R}^{16}$*

and $\mathbf{p}_1, \mathbf{p}_2, \mathbf{p}_3, \mathbf{p}_4 \in \mathbb{R}^4$ *be the four-dimensional components of* $\boldsymbol{p}$ *such that*

$$\boldsymbol{p} = \begin{pmatrix} \mathbf{p}_1 \\ \mathbf{p}_2 \\ \mathbf{p}_3 \\ \mathbf{p}_4 \end{pmatrix}.$$

Then $(-\lambda)$ *is also an eigenvalue of* A_p *with the eigenvector* $\hat{\boldsymbol{p}}$ *in the form*

$$\hat{\boldsymbol{p}} = \begin{pmatrix} \mathbf{p}_2 \\ \mathbf{p}_1 \\ -\mathbf{p}_4 \\ -\mathbf{p}_3 \end{pmatrix}.$$

Proof. The statement can be checked by direct calculation or can be proved the same way as Proposition 3.1. □

As in the case of elastic materials (see 3.14), the general solution $\boldsymbol{s}_p(x_3)$ after extracting real solutions can be expressed in the form:

$$\boldsymbol{s}_p(x_3) = \sum_{j=1}^{16} c_j \boldsymbol{p}^j(x_3), \tag{3.33}$$

where $\boldsymbol{p}^j$ are known real-valued vector functions with images in $\mathbb{R}^{16}$, either exponential or oscillating (see (3.13) for the detailed form of $\boldsymbol{p}^j$). Extracting the components of s_p corresponding to the amplitude functions $\boldsymbol{a}$, $\boldsymbol{b}$, α and β, we obtain the following expression for the solution pair $(\boldsymbol{u}, \phi)$ with the unknown coefficients $\{c_j\}_{j=1}^{16}$:

$$\begin{cases} \boldsymbol{u}(x_1, x_3) = \sum_{j=1}^{16} c_j \boldsymbol{f}^j(x_3) \cdot \cos(\kappa x_1 - \omega t) + \sum_{j=1}^{16} c_j \boldsymbol{g}^j(x_3) \cdot \sin(\kappa x_1 - \omega t), \\ \phi(x_1, x_3) = \sum_{j=1}^{16} c_j \mu^j(x_3) \cdot \cos(\kappa x_1 - \omega t) + \sum_{j=1}^{16} c_j \nu^j(x_3) \cdot \sin(\kappa x_1 - \omega t), \end{cases} \tag{3.34}$$

where real functions $\boldsymbol{f}^j, \boldsymbol{g}^j, \mu^j$, and ν^j are the components of $\boldsymbol{p}_p^j$ corresponding to $\boldsymbol{a}, \boldsymbol{b}, \alpha$, and β, respectively. The representation (3.34) is the piezoelectric analog of (3.15).

Remark 3.5. The approach used in this section to treat piezoelectric materials can be applied without changes to dielectric materials. Dielectric materials are described by the same equations with the coupling tensor e set to zero.

3.4.2 Piezoelectric multi-layered structure

Consider now a structure consisting of N stacked piezoelectric layers. In order to find plane wave solutions in such a structure we proceed the same way as described in Subsection 3.3.3 for elastic layers, except that we take the system (3.32) instead of (3.10) and the representation (3.34) instead of (3.15). This makes the number of unknown coefficients for each layer 16 instead of 12. On the other hand, the set of conditions is enriched by the conditions expressing the electrical consistence at interfaces and free surfaces. Eventually, as will be shown below, the number of unknowns and the number of conditions remain balanced.

Mechanical Conditions. The mechanical conditions at interfaces and free surfaces are described by the same systems (3.22)–(3.24). Though, it should be emphasized here that the stress tensor σ in these relations must be calculated by the formula (3.25), i.e. with taking into account the contribution of the piezoelectric effect. Thus the relations

$$\begin{aligned} \boldsymbol{u}^{(m-1)} &= \boldsymbol{u}^{(m)} && \text{at } x_3 = \tau_{m-1}, \\ \sigma^{(m-1)}\boldsymbol{n} &= \sigma^{(m)}\boldsymbol{n} && \text{at } x_3 = \tau_{m-1} \end{aligned}$$

take the form (compare with (3.19))

$$\begin{aligned} \left(u_i^{(m-1)} - u_i^{(m)}\right)\Big|_{x_3=\tau_{m-1}} &= 0, \quad i = 1,2,3, \\ \left(G_{i3kl}^{(m-1)}\frac{\partial u_k^{(m-1)}}{\partial x_l} + e_{ki3}^{(m-1)}\frac{\partial \phi^{(m-1)}}{\partial x_k} - G_{i3kl}^{(m)}\frac{\partial u_k^{(m)}}{\partial x_l} - e_{ki3}^{(m)}\frac{\partial \phi^{(m)}}{\partial x_k}\right)\Bigg|_{x_3=\tau_{m-1}} &= 0, \quad i = 1,2,3. \end{aligned}$$

Substituting the expressions for $\boldsymbol{u}$ and ϕ from (3.34) and equating the coefficients of $\sin(\kappa x_1 - \omega t)$ and $\cos(\kappa x_1 - \omega t)$, we obtain 12 linear equations for unknown coefficients

$$\{c_j^{(m-1)}\}_{j=1}^{16}, \ \{c_j^{(m)}\}_{j=1}^{16}.$$

This forms 12 rows in the fitting matrix F for each pair of neighboring layers $(m-1, m)$. The mechanical condition on the free surface of the top (N-th) layer takes the forms

$$\left(G_{i3kl}^{(N)}\frac{\partial u_k^{(N)}}{\partial x_l} + e_{ki3}^{(N)}\frac{\partial \phi^{(N)}}{\partial x_k}\right)\Bigg|_{x_3=\tau_N} = 0, \quad i = 1,2,3. \tag{3.35}$$

This condition contributes 6 more raws in the fitting matrix.

Electrical Conditions. The electrical interface conditions express the continuity of the normal component of the electric displacement ($\boldsymbol{D} \cdot \boldsymbol{n} = D_3$) and the continuity of the tangent component of the electric field. Hence at the interface between layers $(m-1)$ and m we have

$$D_3^{(m-1)} = D_3^{(m)} \qquad \text{at } x_3 = \tau_{m-1},$$
$$E_1^{(m-1)} = E_1^{(m)} \qquad \text{at } x_3 = \tau_{m-1}.$$

Substituting $\boldsymbol{D}$ and $\boldsymbol{E}$ from (3.26) and (3.27), we obtain

$$\begin{aligned} \left(\epsilon_{3j}^{(m-1)} \frac{\partial \phi^{(m-1)}}{\partial x_j} + e_{3kl}^{(m-1)} \frac{\partial u_l^{(m-1)}}{\partial x_k} - \epsilon_{3j}^{(m)} \frac{\partial \phi^{(m)}}{\partial x_j} - e_{3kl}^{(m)} \frac{\partial u_l^{(m)}}{\partial x_k} \right) \Bigg|_{x_3=\tau_{m-1}} &=0, \\ \left(\frac{\partial \phi^{(m-1)}}{\partial x_1} - \frac{\partial \phi^{(m)}}{\partial x_1} \right) \Bigg|_{x_3=\tau_{m-1}} &=0. \end{aligned} \tag{3.36}$$

Note that we do not equate the x_2 components of the electric field E because the electric potential ϕ does not depend on x_2. This means that $E_2 = 0$ and this condition is fulfilled automatically.

The electrical conditions on the free surface depend on the contacting medium. Here we assume that no electrical field may exist in it (Section 3.5 considers more realistic conditions). In this case, the following relations on the free surface of the top (and/or bottom) layer hold:

$$\begin{aligned} \left(\epsilon_{3j}^{(m)} \frac{\partial \phi^{(m)}}{\partial x_j} + e_{3kl}^{(m)} \frac{\partial u_l^{(m)}}{\partial x_k} \right) \Bigg|_{x_3=\tau_m} &=0, \\ \frac{\partial \phi^{(m)}}{\partial x_1} \Bigg|_{x_3=\tau_m} &=0, \end{aligned} \tag{3.37}$$

where $m = N$ for the top surface of the top layer, and $m = 0$ for the bottom surface of the bottom layer.

Combining the mechanical and electrical conditions, we obtain 16 linear equations connecting the unknown coefficients for each pair of neighboring layers. The conditions on a free surface (if present) give 10 scalar equations (6 mechanical + 4 electrical). On the other hand, we have 16 unknown coefficients for each layer. Therefore, if a free surface is present, the number of equations exceeds the number of unknowns by 2 for each free surface. That is, we have even more equations than needed. This imperfectness is fixed in Section 3.5 by considering the electrical influence of the surroundging medium more carefully, which requires the introduction of two more unknown coefficients.

As usual, if a semi-infinite layer is present, the condition on the free surface is replaced with the requirement of the wave decaying. This decreases the number of equations by 10 but removes at least 8 unknowns due to Proposition 3.4. Thus the system of equations remains not underdetermined, and the algorithm proposed remains applicable.

3.4.3 Mixed multi-layered structure

We consider now mixed multi-layered structures consisting of piezoelectric and elastic layers. From the electrical point of view elastic layers can be either conductive or dielectric. In the former case the layer is treated as a pure elastic material as described in Section 3.3 without taking into account electrical phenomena. In the latter case it is described by the same relations as a piezoelectric layer with the piezoelectric tensor e set to zero (see Remark 3.5).

Remark 3.6. From the mathematical point of view, we can treat conductive elastic layers as a special case of piezoelectric ones by setting e and ϕ to zero. The electrical conditions on free surfaces and at interfaces between conductive layers would degenerate to trivial $0 = 0$, the conditions at interfaces with piezoelectric layers would make sense as well. Thus, from the mathematical point of view, a mixed structure could be considered as a piezoelectric multi-layered structure.
However, from the computational point of view, treatment of conductive elastic layers as piezoelectric ones requires redundant calculations and therefore increases the caculation error. Hence they are treated separately.

In mixed structures we have to specify the interface conditions between layers of different types. By Remark 3.5, only conditions at interfaces between conductive elastic and piezoelectric layers require a special treatment.

Let the superscripts 1 and 2 denote the piezoelectric and the conductive layer, respectively. The mechanical conditions at the interface remain the same as above. They consist in the continuity of the displacement field and the pressure equilibrium. The only thing one should keep in mind is that the stress is calculated in different ways. We have

$$u_i^{(1)} = u_i^{(2)}, \qquad i = 1, 2, 3,$$

$$G_{i3kl}^{(1)} \frac{\partial u_k^{(1)}}{\partial x_l} + e_{ki3}^{(1)} \frac{\partial \phi^{(1)}}{\partial x_k} = G_{i3kl}^{(2)} \frac{\partial u_k^{(2)}}{\partial x_l}, \qquad i = 1, 2, 3.$$

In the conductive layer there is no electric field; the electrical potential ϕ is not a relevant variable for this layer. Therefore, the electrical conditions at the interface are like those on a free surface of piezoelectric (see (3.37)), that is,

$$\begin{aligned} \epsilon_{3j}^{(1)} \frac{\partial \phi^{(1)}}{\partial x_j} + e_{3kl}^{(1)} \frac{\partial u_l^{(1)}}{\partial x_k} =& 0, \\ \frac{\partial \phi^{(1)}}{\partial x_1} =& 0. \end{aligned}$$

It can be verified directly that the total number of unknown coefficients does not exceed the number of equations originated from the boundary and interface conditions, does not matter how the layers are mixed. This becomes obvious in view of Remark 3.6.

3.5 Contact with Surrounding Dielectric Medium

As mentioned above the electrical conditions (3.37) on the free surface are seldom realistic because they assume the absence of the electric field outside the solid structure, which is not true as a rule. In this section we assume that the free surface contacts with a dielectric isotropic medium whose mechanical influence is negligible. It can be for example vacuum or gas. Since the medium is isotropic, we assume the following constitutive relation:

$$\boldsymbol{D} = \epsilon_0 \, \boldsymbol{E}, \tag{3.38}$$

where ϵ_0 is the vacuum permittivity. Further, by Gauss's law we have

$$\mathrm{div}\boldsymbol{D} = 0. \tag{3.39}$$

Combining (3.38) and (3.28), we obtain Laplace's equation for the electric potential ϕ, that is,

$$\Delta \phi = 0.$$

As usual, we are looking for plane wave solutions to this equation in the form

$$\phi(x_1, x_3) = \alpha(x_3) \cos(\kappa x_1 - \omega t) + \beta(x_3) \sin(\kappa x_1 - \omega t).$$

Substituting it to the Laplace equation yields two elementary differential equations for α and β. Solving them, we obtain

$$\phi(x_1, x_3) = \left(c_1 e^{-\kappa x_3} + c_2 e^{\kappa x_3}\right)\cos(\kappa x_1 - \omega t) + \left(c_3 e^{-\kappa x_3} + c_4 e^{\kappa x_3}\right)\sin(\kappa x_1 - \omega t),$$

where c_1, c_2, c_3 and c_4 are arbitrary coefficients. Note that only two of them are not zero, since only solutions decaying as $x_3 \to +\infty$ (or $x_3 \to -\infty$ if the contacting medium is located under the structure) are left.

The electrical conditions (3.37) at the top surface are corrected now as follows:

$$\begin{aligned} \left(\epsilon_{3j}^{(N)}\frac{\partial \phi^{(N)}}{\partial x_j} + e_{3kl}^{(N)}\frac{\partial u_l^{(N)}}{\partial x_k}\right)\bigg|_{x_3=\tau_N} &= \epsilon_0 \frac{\partial \phi}{\partial x_3}\bigg|_{x_3=\tau_N}, \\ \frac{\partial \phi^{(N)}}{\partial x_1}\bigg|_{x_3=\tau_N} &= \frac{\partial \phi}{\partial x_1}\bigg|_{x_3=\tau_N}, \end{aligned} \tag{3.40}$$

where N is the index of the top layer. These modified conditions involve two additional unknown coefficients. However, this does not make the fitting matrix underdetermined because with the uncorrected conditions (3.37) the system had two redundant equations (see Subsection 3.4.2).

3.6 Contact with Fluid

Assume now that the upper surface of the top layer contacts with a viscous compressible fluid. Its motion is described by the Navier-Stokes equations

$$\begin{aligned} \varrho\,(\boldsymbol{v}_t + (\boldsymbol{v}\cdot\nabla)\boldsymbol{v}) &= -\,\nabla p + \nu\Delta\boldsymbol{v} + (\zeta + \frac{\nu}{3})\nabla(\operatorname{div}\boldsymbol{v}), \\ \varrho_t &= -\operatorname{div}(\varrho\boldsymbol{v}), \end{aligned} \tag{3.41}$$

where $\boldsymbol{v}$ is the velocity field, p is the pressure, $\nu > 0$ and $\zeta > 0$ are the dynamic and volume viscosities, respectively; ϱ is the density of the fluid. The stress tensor σ is determined as follows:

$$\sigma_{ik} = -p\delta_{ik} + \nu\left(\frac{\partial v_i}{\partial x_k} + \frac{\partial v_k}{\partial x_i}\right) + (\zeta - \frac{2}{3}\nu)\delta_{ik}\operatorname{div}\boldsymbol{v}. \tag{3.42}$$

The oscillations of fluid particles in acoustic waves are small (see [41]). This enables us to neglect the non-linear term $(\boldsymbol{v}\cdot\nabla)\boldsymbol{v}$ on the left-hand side. For the same reason, we assume that the relative changes in the density and the pressure are small and admit the form

$$\varrho = \varrho_0 + \varrho', \quad p = p_0 + p',$$

where ϱ_0 and p_0 are the constant static density and pressure, respectively; ϱ' and p' are their changes in the acoustic wave. We assume that

$$\varrho' \ll \varrho_0, \quad p' \ll p_0.$$

Furthermore, we assume ϱ' and p' to be related linearly, that is,

$$\varrho' = \alpha p',$$

where α is a constant expressing the compressibility of the fluid. We have then

$$\varrho(p') = \varrho_0 + \alpha p'.$$

Assuming ϱ', p', and $\boldsymbol{v}$ to be small and neglecting all terms of the second order of smallness, we can rewrite (3.41) as follows:

$$\begin{aligned} \varrho_0 \boldsymbol{v}_t + \nabla p - \nu \Delta \boldsymbol{v} - (\zeta + \frac{\nu}{3}) \nabla (\operatorname{div} \boldsymbol{v}) &= 0, \\ \alpha p_t + \varrho_0 \operatorname{div} \boldsymbol{v} &= 0. \end{aligned} \tag{3.43}$$

Similar to the case of solid structures, we are looking for plain wave solutions in the form

$$\begin{aligned} \boldsymbol{v}(x_1, x_3) &= \boldsymbol{a}(x_3) \cos(\kappa x_1 - \omega t) + \boldsymbol{b}(x_3) \sin(\kappa x_1 - \omega t), \\ p(x_1, x_3) &= f(x_3) \cos(\kappa x_1 - \omega t) + g(x_3) \sin(\kappa x_1 - \omega t). \end{aligned}$$

Substituting this representation into the second equation in (3.43), we can express the amplitudes f and g through $\boldsymbol{a}$ and $\boldsymbol{b}$ as follows:

$$\begin{aligned} f(x_3) &= \frac{\varrho_0 \kappa}{\alpha \omega} [a_1(x_3) - \dot{b}_3(x_3)], \\ g(x_3) &= \frac{\varrho_0 \kappa}{\alpha \omega} [b_1(x_3) + \dot{a}_3(x_3)], \end{aligned}$$

where the dot denotes the differentiation with respect to $\tilde{x}_3 = \kappa x_3$. The equations (3.43) yield then the system only for the amplitude functions $\boldsymbol{a}$ and $\boldsymbol{b}$. The system reads:

$$\left(\begin{array}{ccc|ccc} -\nu & 0 & 0 & 0 & 0 & 0 \\ 0 & -\nu & 0 & 0 & 0 & 0 \\ 0 & 0 & -\mu & 0 & 0 & -\gamma \\ \hline 0 & 0 & 0 & \nu & 0 & 0 \\ 0 & 0 & 0 & 0 & \nu & 0 \\ 0 & 0 & -\gamma & 0 & 0 & \mu \end{array}\right)\begin{pmatrix} \ddot{a}_1 \\ \ddot{a}_2 \\ \ddot{a}_3 \\ \ddot{b}_1 \\ \ddot{b}_2 \\ \ddot{b}_3 \end{pmatrix} = \left(\begin{array}{ccc|ccc} 0 & 0 & -\gamma & 0 & 0 & \eta \\ 0 & 0 & 0 & 0 & 0 & 0 \\ -\gamma & 0 & 0 & \eta & 0 & 0 \\ \hline 0 & 0 & \eta & 0 & 0 & \gamma \\ 0 & 0 & 0 & 0 & 0 & 0 \\ \eta & 0 & 0 & \gamma & 0 & 0 \end{array}\right)\begin{pmatrix} \dot{a}_1 \\ \dot{a}_2 \\ \dot{a}_3 \\ \dot{b}_1 \\ \dot{b}_2 \\ \dot{b}_3 \end{pmatrix} + \tag{3.44}$$

$$+\left(\begin{array}{ccc|ccc} -\mu & 0 & 0 & \varrho_0\omega/k^2-\gamma & 0 & 0 \\ 0 & -\nu & 0 & 0 & \varrho_0\omega/k^2 & 0 \\ 0 & 0 & -\nu & 0 & 0 & \varrho_0\omega/k^2 \\ \hline \varrho_0\omega/k^2-\gamma & 0 & 0 & \mu & 0 & 0 \\ 0 & \varrho_0\omega/k^2 & 0 & 0 & \nu & 0 \\ 0 & 0 & \varrho_0\omega/k^2 & 0 & 0 & \nu \end{array}\right)\begin{pmatrix} a_1 \\ a_2 \\ a_3 \\ b_1 \\ b_2 \\ b_3 \end{pmatrix}.$$

Here $\mu := \zeta + \frac{4\nu}{3}$, $\gamma := \frac{\varrho_0}{\alpha\omega}$, and $\eta := \zeta + \frac{\nu}{3}$. Denote the matrices of the system by M, K, L so that it takes the form

$$M\begin{pmatrix} \ddot{\boldsymbol{a}} \\ \ddot{\boldsymbol{b}} \end{pmatrix} = K\begin{pmatrix} \dot{\boldsymbol{a}} \\ \dot{\boldsymbol{b}} \end{pmatrix} + L\begin{pmatrix} \boldsymbol{a} \\ \boldsymbol{b} \end{pmatrix}.$$

Remark. The matrix M is non-singular. Indeed,

$$\det M = (-\nu)\cdot(-\nu)\cdot\nu\cdot\nu\cdot(-\mu^2-\lambda^2) = -\nu^4(\mu^2+\lambda^2) \neq 0.$$

The matrix M is thus invertible. This enables us to rewrite the system (3.44) in the normal form as follows:

$$\dot{\boldsymbol{s}}_f = A_f \boldsymbol{s}_f, \tag{3.45}$$

where

$$\boldsymbol{s}_f := \begin{pmatrix} \boldsymbol{a} \\ \boldsymbol{b} \\ \dot{\boldsymbol{a}} \\ \dot{\boldsymbol{b}} \end{pmatrix} \in \mathbb{R}^{12}$$

is the state vector, and

$$A_f := \begin{pmatrix} 0 & \mathbf{I}_6 \\ M^{-1}L & M^{-1}K \end{pmatrix}$$

is the matrix of the system.

Proposition 3.7. *Let λ be an eigenvalue of the matrix A_f. Then $(-\lambda)$ is also an eigenvalue of A_f.*

Proof. This proposition is the analog of Propositions (3.1) and (3.4). The proof is quite straightforward. The characteristic matrix of A_f satisfies

$$(A_f - \lambda \mathbf{I}_{12}) = \begin{pmatrix} -\lambda \mathbf{I}_6 & \mathbf{I}_6 \\ M^{-1}L & M^{-1}K - \lambda \mathbf{I}_6 \end{pmatrix} = \begin{pmatrix} \mathbf{I}_6 & 0 \\ 0 & M^{-1} \end{pmatrix} \begin{pmatrix} -\lambda \mathbf{I}_6 & \mathbf{I}_6 \\ L & K - \lambda M \end{pmatrix}$$

The characteristic equation is then reduced to

$$\det \begin{pmatrix} -\lambda \mathbf{I}_6 & \mathbf{I}_6 \\ L & K - \lambda M \end{pmatrix} = 0.$$

This is equivalent to

$$\det \begin{pmatrix} 0_6 & \mathbf{I}_6 \\ L + \lambda(K - \lambda M) & K - \lambda M \end{pmatrix} = 0 \quad \Leftrightarrow \quad \det(L + \lambda K - \lambda^2 M) = 0.$$

The matrix $L + \lambda K - \lambda^2 M$ is

$$\left(\begin{array}{ccc|ccc} -\mu + \lambda^2 \nu & 0 & -\lambda\gamma & \varrho_2 & 0 & \lambda\eta \\ 0 & -\nu + \lambda^2 \nu & 0 & 0 & \varrho_1 & 0 \\ -\lambda\gamma & 0 & -\nu + \lambda^2 \mu & \lambda\eta & 0 & \varrho_1 + \lambda^2\gamma \\ \hline \varrho_2 & 0 & \lambda\eta & \mu - \lambda^2\nu & 0 & \lambda\gamma \\ 0 & \varrho_1 & 0 & 0 & \nu - \lambda^2 \nu & 0 \\ \lambda\eta & 0 & \varrho_1 + \lambda^2\gamma & \lambda\gamma & 0 & \nu - \lambda^2\mu \end{array}\right)$$

where $\varrho_1 := \varrho_0\omega/k^2$; $\varrho_2 := \varrho_0\omega/k^2 - \gamma$. Rearranging raws and columns, we obtain

$$\left(\begin{array}{cc|cc|cc} \lambda\eta & -\lambda\gamma & \varrho_2 & -\mu + \lambda^2\nu & 0 & 0 \\ \lambda\gamma & \lambda\eta & \mu - \lambda^2\nu & \varrho_2 & 0 & 0 \\ \hline \varrho_1 + \lambda^2\gamma & -\nu + \lambda^2\mu & \lambda\eta & -\lambda\gamma & 0 & 0 \\ \nu - \lambda^2\mu & \varrho_1 + \lambda^2\gamma & \lambda\gamma & \lambda\eta & 0 & 0 \\ \hline 0 & 0 & 0 & 0 & \varrho_1 & (\lambda^2 - 1)\nu \\ 0 & 0 & 0 & 0 & \nu(1 - \lambda^2) & \varrho_1 \end{array}\right).$$

Te characteristic equation is satisfied iff

$$\det\left(\begin{array}{cc|cc} \lambda\eta & -\lambda\gamma & \varrho_2 & -\mu+\lambda^2\nu \\ \lambda\gamma & \lambda\eta & \mu-\lambda^2\nu & \varrho_2 \\ \hline \varrho_1+\lambda^2\gamma & -\nu+\lambda^2\mu & \lambda\eta & -\lambda\gamma \\ \nu-\lambda^2\mu & \varrho_1+\lambda^2\gamma & \lambda\gamma & \lambda\eta \end{array}\right) = 0.$$

This yields a polynomial equation with respect to λ. It can be seen that the polynomial contains no terms of odd degree. That is, the equation is of the form

$$k_8\lambda^8 + k_6\lambda^6 + k_4\lambda^4 + k_2\lambda^2 + k_0 = 0.$$

The statement of the proposition becomes now obvious. Indeed, if λ satisfies the characteristic equation, $(-\lambda)$ satisfies it as well. □

We treat now the system (3.45) the same way as its elastic and piezoelectric counterparts in Sections 3.3 and 3.4. The general solution can be expressed in the form

$$\boldsymbol{s}_f(x_3) = \sum_{j=1}^{12} c_j \boldsymbol{p}^j(x_3), \tag{3.46}$$

where $\boldsymbol{p}^j$ are known real-valued vector functions, either exponential or oscillating (see (3.13) for details). Extracting the components of $\boldsymbol{s}_f$ corresponding to the amplitude functions $\boldsymbol{a}$ and $\boldsymbol{b}$, we obtain the following expression for $\boldsymbol{v}$ with the unknown coefficients $\{c_j\}_{j=1}^{12}$.

$$\boldsymbol{v}(x_1, x_3) = \sum_{j=1}^{12} c_j \boldsymbol{f}^j(x_3)\cdot\cos(\kappa x_1 - \omega t) + \sum_{j=1}^{12} c_j \boldsymbol{g}^j(x_3)\cdot\sin(\kappa x_1 - \omega t). \tag{3.47}$$

Here real functions $\boldsymbol{f}^j$ and $\boldsymbol{g}^j$ are the components of $\boldsymbol{p}^j$ corresponding to $\boldsymbol{a}$ and $\boldsymbol{b}$ respectively.

Similar to the case of semi-infinite elastic/piezoelectric materials we require the amplitudes to decay when moving away from the surface. Proposition 3.7 ensures that at least a half of summands in (3.46) disappears due to this requirement so that no more than 6 summands are left.

Conditions on the solid-fluid interface. The matching mechanical conditions at the interface between the fluid and the top layer consist in the continuity of the velocities and the pressure equilibrium. Suppose that the top layer contacting the fluid is piezoelectric and has the index N. Then the conditions take the form

$$\frac{\partial u_i^{(N)}}{\partial t} = v_i, \tag{3.48}$$

$$C_{i3kl}^{(N)} \frac{\partial u_l^{(N)}}{\partial x_k} + e_{ki3}^{(N)} \frac{\partial \phi^{(N)}}{\partial x_k} = -\, p\delta_{i3} + \nu \left(\frac{\partial v_i}{\partial x_3} + \frac{\partial v_3}{\partial x_i} \right) + (\zeta - \frac{2}{3}\nu)\delta_{i3} \operatorname{div} \boldsymbol{v}. \tag{3.49}$$

The pressure equilibrium is the same condition as at interfaces between layers, only rewritten with the stress tensor for the fluid determined by (3.42). The continuity of the velocities condition couples the oscillations in the solid and the fluid. Note that if the top layer is not piezoelectric, the term with e on the left-hand side of (3.49) disappears.

Additional to the mechanical conditions we have to specify the proper electrical conditions at the contact interface. If the top layer is conductive, we can omit them because the electric field vanishes in it, and the electrical properties of the fluid have no influences on the rest of the structure. Otherwise, we treat the electrical influence of the fluid exactly the same way as that of the dielectric medium (see Section 3.5, (3.40)). That is, we have the following electric conditions:

$$\epsilon_{3j}^{(N)} \frac{\partial \phi^{(N)}}{\partial x_j} + e_{3kl}^{(N)} \frac{\partial u_l^{(N)}}{\partial x_k} = \epsilon_0 \frac{\partial \phi}{\partial x_3}, \tag{3.50}$$

$$\frac{\partial \phi^{(N)}}{\partial x_1} = \frac{\partial \phi}{\partial x_1}, \tag{3.51}$$

where ϵ_0 is the dielectric permittivity; ϕ is the electric potential of the fluid.

In contrast to the case of contact with a dielectric medium discussed in Section 3.5 we introduce here an additional variable $\boldsymbol{v}$, which increases the total number of unknown coefficients by 6 (see 3.46). On the other hand we have the additional condition (3.48) that yields 6 more equations. Thus the number of equations and unknowns remains in balance.

3.7 Bristle-Like Structure at Fluid-Solid Interface

In this section we consider again a solid contacting with a fluid, but now we assume that the contacting surface of the solid is covered with a very thin and dense periodic bristle-like structure (see Figure 3.4). The necessity of the investigation of such structures arises for example when modeling the biosensor described in Section 1.1, which was the initial motivation of the work.

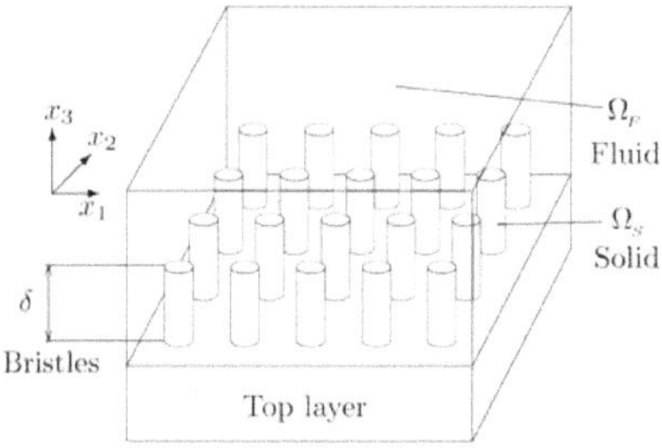

Figure 3.4: Bristle-like solid-fluid interface.

We assume the height of the bristles to be very small and their density to be very high. The direct modeling of such a structure using fluid-solid interface conditions is impossible. Instead of that, we exploit the homogenization technique developed in [28]. We briefly reproduce here the main results.

The main idea is to replace the bristle-like interlayer by an averaged layer whose properties are derived as the number of bristles goes to infinity whereas their thickness goes to zero, thereby the height of the bristles remains constant.

Let Ω_S be the domain occupied by the bristles and Ω_F be the domain occupied by the fluid between the bristles. Denote by Γ the interface between Ω_S and Ω_F. The whole domain Ω occupied by the layer is then $\Omega_S \cup \Gamma \cup \Omega_F$. The structure is assumed to be periodic in the x_1- and x_2-directions and independent on x_3. The basic governing equations read as follows:

$$\varrho_F \boldsymbol{v}_t + \nabla p - \operatorname{div}(P\nabla \boldsymbol{v}) = 0 \quad \text{in } \Omega_F, \tag{3.52}$$

$$\alpha p_t + \varrho_F \operatorname{div} \boldsymbol{v} = 0 \quad \text{in } \Omega_F, \tag{3.53}$$

$$\varrho_S \boldsymbol{u}_{tt} - \operatorname{div}(G\nabla \boldsymbol{u}) = 0 \quad \text{in } \Omega_S. \tag{3.54}$$

These are the same equation as (3.43) and (3.6) rewritten in tensor form; ϱ_F and ϱ_S are the densities of the fluid and the solid parts, respectively. The fourth-rank tensors P expresses the fluid viscosity and is determined by

$$P\nabla \boldsymbol{v} = \nu \left(\nabla \boldsymbol{v} + \nabla^T \boldsymbol{v}\right) + (\zeta - \frac{2}{3}\nu)\, I \operatorname{div} \boldsymbol{v},$$

where I is the unit tensor with components $I_{ij} = \delta_{ij}$.

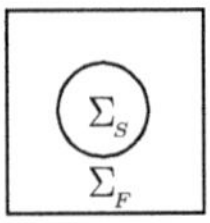

Figure 3.5: Periodic cell Σ.

The no-slip and pressure equilibrium conditions on the fluid-solid interface read:

$$\boldsymbol{u}_t = \boldsymbol{v} \qquad \text{on} \quad \Gamma, \tag{3.55}$$

$$G\nabla\boldsymbol{u}\cdot\boldsymbol{n} = (-pI + P\nabla\boldsymbol{v})\cdot\boldsymbol{n} \qquad \text{on} \quad \Gamma. \tag{3.56}$$

The condition (3.55) is the main difficulty for the mathematical treatment of the model (3.52)–(3.56). In order to overcome it, the approach proposed by J.-L. Lions in [46] is used. The basic idea is to use the velocity instead of the displacement in equation (3.54) as the state variable. This is done by introducing the following integral operator:

$$\mathcal{J}_t\boldsymbol{v} := \int_0^t \boldsymbol{v}(s)ds.$$

The equation (3.54) takes then the form

$$\varrho_S\boldsymbol{v}_t - \operatorname{div} G\mathcal{J}_t\nabla\boldsymbol{v} = 0, \tag{3.57}$$

where $\boldsymbol{v} = \boldsymbol{u}_t$. The pressure p in (3.53) is expressed through the velocity $\boldsymbol{v}$ as follows:

$$p = -\frac{\varrho_F}{\alpha}\operatorname{div}\mathcal{J}_t\boldsymbol{v}. \tag{3.58}$$

Assume that the (x_1, x_2)-projection of the periodic cell of the bristle structure is a square containing just one bristle (see Figure 3.5). Denote it by Σ. Let Σ_S be the projection of the bristle, and $\Sigma_F = \Sigma \setminus \overline{\Sigma_S}$. Further, let $\hat{\chi}(x_1, x_2)$ be the Σ-periodic extension of the characteristic function of the domain Σ_F to $\mathbb{R}^2$. We define then the characteristic function of the domain occupied by the fluid as follows:

$$\chi^\varepsilon(\boldsymbol{x}) = \hat{\chi}\left(\frac{x_1}{\varepsilon}, \frac{x_2}{\varepsilon}\right), \tag{3.59}$$

where ε is a refinement parameter. The value $\varepsilon = 1$ corresponds to the original structure; the bristles become finer and their density grows whenever $\varepsilon \to 0$.

Using (3.59), (3.57), and (3.58), we can rewrite the original equations (3.52) – (3.54) as one equation with discontinues coefficients in the whole domain Ω as follows:

$$\varrho^{\varepsilon}\, \boldsymbol{v}_t^{\varepsilon} - \operatorname{div}\left(M^{\varepsilon}\nabla \boldsymbol{v}^{\varepsilon}\right) = 0 \qquad \text{in } \Omega, \tag{3.60}$$

where

$$\varrho^{\varepsilon} := \varrho_F \chi^{\varepsilon} + \varrho_S(1-\chi^{\varepsilon}),$$
$$M^{\varepsilon} := \chi^{\varepsilon} P + \left(\chi^{\varepsilon}\frac{\varrho_F}{\alpha} I \otimes I + (1-\chi^{\varepsilon})G\right)\mathcal{J}_t.$$

The interface condition (3.55) is equivalent to the continuity of $\boldsymbol{v}^{\varepsilon}$ on Γ; the condition (3.56) takes now the form

$$G\mathcal{J}_t\nabla \boldsymbol{v}^{\varepsilon}\cdot \boldsymbol{n} = \left(\frac{\varrho_F}{\alpha}\operatorname{div}\mathcal{J}_t \boldsymbol{v}^{\varepsilon}\cdot I + P\nabla \boldsymbol{v}^{\varepsilon}\right)\cdot \boldsymbol{n} \quad \text{on } \Gamma^{\varepsilon}. \tag{3.61}$$

The problem (3.60)–(3.61) is treated then by the two-scale method. The homogenized material is described by the following limiting equation (see [28] for details):

$$\varrho_\theta \boldsymbol{v}_t - \operatorname{div}\mathcal{J}_t\hat{G}\nabla \boldsymbol{v} - \operatorname{div}\hat{P}\nabla \boldsymbol{v} - \operatorname{div}\int_0^t \omega(t-s)\nabla \boldsymbol{v}(s)\,ds = 0. \tag{3.62}$$

The term containing the tensor $\hat{P}$ describes the viscous damping and originates from the fluid part. The term containing the tensor $\hat{G}$ represents elastic stresses. The tensor $\hat{G}$ is degenerate, its kernel is such that volume preserving deformations (shear deformations in particular) do not produce elastic stresses. The integral term represents a memory effect that is responsible for viscoelastic properties of the limiting material. It is stated numerically that the memory effect is very weak. The system "forgets" the current history very quickly. Therefore, the integral term on the left-hand-side of (3.62) can be dropped. Note that the limiting material described by (3.62) is as a rule anisotropic even if the material of the solid part of the bristle structure is isotropic. This occurs because the limiting material inherits geometric properties of the bristle structure. The computation of the tensors $\hat{P}$, $\hat{G}$, and $\omega(\tau)$ is based on an analytical representation of solutions of the so-called cell equation arising in homogenization theory. The numerical calculation of them involves the finite element method.

Substituting $\boldsymbol{u} = \mathcal{J}_t\boldsymbol{v}$ and dropping the integral, we obtain the final equation for bristle-like layers in the form

$$\varrho \boldsymbol{u}_{tt} - \operatorname{div}\left(\hat{G}\nabla \boldsymbol{u}\right) - \operatorname{div}\left(\hat{P}\nabla \boldsymbol{u}_t\right) = 0. \tag{3.63}$$

The conditions at the interface with the underlying solid layer and the overlying fluid are similar to those described in Sections 3.3 and 3.6, respectively. The only thing to take into account is that the stress vector acting on the surface with the normal vector $\boldsymbol{n}$ is expressed by

$$\sigma \cdot \boldsymbol{n} = \left(\hat{G}\nabla \boldsymbol{u} + \hat{P}\nabla \boldsymbol{u}_t\right) \cdot \boldsymbol{n}. \tag{3.64}$$

Note that all the terms in (3.63) and (3.64) are linear with respect to $\boldsymbol{u}$. This enables a straightforward integration of such layers into the general scheme.

3.8 Introducing Multilayers

In this section we consider the propagation of plain acoustic waves in a material consisting of a series of periodically alternating sublayers (see Figure 3.6). In the physical literature, such structures are sometimes referred to as multilayers. In the last time they are intensively investigated by physicists due to a number of very promising properties.

The sublayers are very thin, their thickness may be just a few nanometers, which is significantly less than the wavelength. This fact and a large number of sublayers make the direct modeling of such multi-layered structures impossible. Therefore, instead of this, we replace the original composite material with an averaged one. The exploited homogenization technique is described in details in Chapter 4. Here we only use the limiting relations derived there.

We assume that the periodic set consists of M sublayers. Denote by h_s the relative thickness of the s-th sublayer such that:

$$\sum_{s=1}^{M} h_s = 1.$$

The stiffness tensor of the s-th sublayer denote by G^s.

As established in Chapter 4, the homogenized material is described by the same linear elasticity equation with a constant effective stiffness tensor. Denote the stiffness tensor of

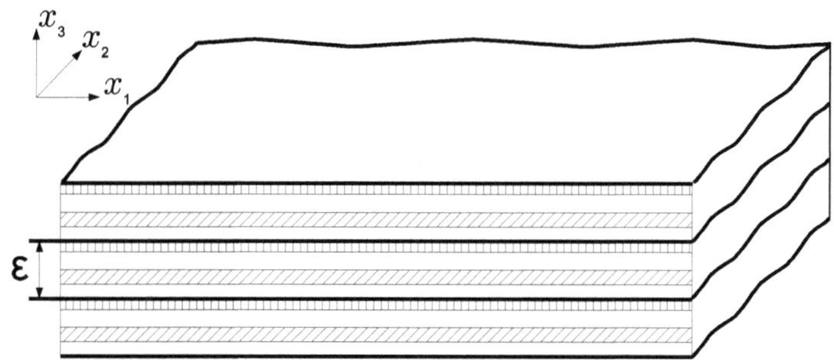

Figure 3.6: Multilayers.

the homogenized material by G^{hom}. By Theorem (4.13) from Section 4.4, we have

$$G^{\text{hom}}_{mnkl} = \sum_{s=1}^{M} h_s C^s_{mnkl} + \sum_{s=1}^{M} h_s C^s_{i3kl} \widehat{N}^s_{imn},$$

where $\widehat{N}^s_{imn}$ are auxiliary variables determined by

$$\widehat{N}^s_{imn} = (G^s_{\cdot 3 \cdot 3})^{-1}_{iq} \left[\left(\sum_{r=1}^{M} h_r (G^r_{\cdot 3 \cdot 3})^{-1} \right)^{-1}_{pq} \sum_{r=1}^{M} h_r (G^r_{\cdot 3 \cdot 3})^{-1}_{pj} G^r_{mnj3} - G^s_{mnq3} \right].$$

Moreover, it is proved in Theorem 4.9 that the limiting tensor G^{hom} possesses the same nice properties as an ordinary stiffness tensor. That is, it is positive definite and symmetric. These facts enable us to treat the homogenized multilayers exactly the same way as ordinary elastic layers as described in Section 3.3. It is however useful and convenient to enrich the computer implementation of the model by integrating a tool for the explicit computation of the parameters of homogenized multilayers, because elastic properties of multilayers can be interesting on their own account, regardless of acoustic waves (see e.g. [25]).

3.9 Constructing Dispersion Curves

This section describes a method for building dispersion relations, that is, relations between the phase velocity and the exciting frequency. Algorithm 3.1 presented in Section 3.3 allows to determine the phase velocity of an acoustic wave feasible in a given structure at a given frequency. This yields a single point lying on the dispersion curve corresponding to the found wave mode. In order to build the curve for a range of frequencies the algorithm has to be run many times for different values of the frequency. Doing this by hand is boring because every single calculation involves the choice of the interval on which the fitting function is to minimize. Besides, it must be ensured that the velocity remains on the dispersion curve corresponding to the same wave mode and does not jump to other possible wave modes as the frequency changes. In order to automate this boring job we developed a simple tool that tries to extend the found part of the curve to a wider range of frequencies. It works as follows.

Denote the ordered collection of the found points by $\{(\omega_1, v_1), (\omega_2, v_2), ..., (\omega_n, v_n)\}$ such that $\omega_i > \omega_j$ whenever $i > j$. Suppose we would like to extend the curve to the interval $[\omega_n, \omega_{n+m}]$ with the step $\Delta\,\omega$ such that $\omega_{n+j} := \omega_n + j\ \Delta\,\omega, 0 < j \leqslant m$. Assume for the time being that $n \geqslant 2$. As the model states, the phase velocity at the frequency ω_{n+1} is a minimizer of the fitting function calculated at this frequency. Since the fitting function is usually not convex and may have several minimizers, the minimization interval should be small enough to exclude non-relevant minimizers. At the same time, it must contain the wanted one. We construct it as follows. First, on the base of the last two points, we build the linear extrapolation of the curve (see Figure 3.9). Its value at the point ω_{n+1} denoted by $\overline{v}$ is determined by:

$$\overline{v} := \frac{v_{n-1} - v_n}{\omega_{n-1} - \omega_n}\omega_{n+1} + \frac{\omega_n v_{n-1} - \omega_{n-1} v_n}{\omega_{n-1} - \omega_n}$$

We set then the minimization interval to $[\overline{v} - d_l, \overline{v} + d_r]$, where d_l and d_r are non-negative parameters of the algorithm specified by the user. The minimizer of fitting function on this interval is taken as v_{n+1}. As the point (ω_{n+1}, v_{n+1}) is constructed we can proceed with the construction of the point (ω_{n+2}, v_{n+2}) and so on.

Consider now the case $n = 1$. In this case we possess only one point on the curve and therefore can not construct a linear approximation, so we take a constant approximation, i.e. we take $\overline{v} := v_1$.

The algorithm proposed here is quite heuristic, it relies on the right choice of the parameters d_l and d_r that form the minimization interval. It is expected that the real minimizer does not deviate much from the linear approximation. Though, if the expectation interval is too small, it may happen that the minimizer lies outside of it. In this case the minimum value is most likely found at the start or at the end of the interval. This allows to detect such situations and may suggest to increase the interval. A worse situation may take place

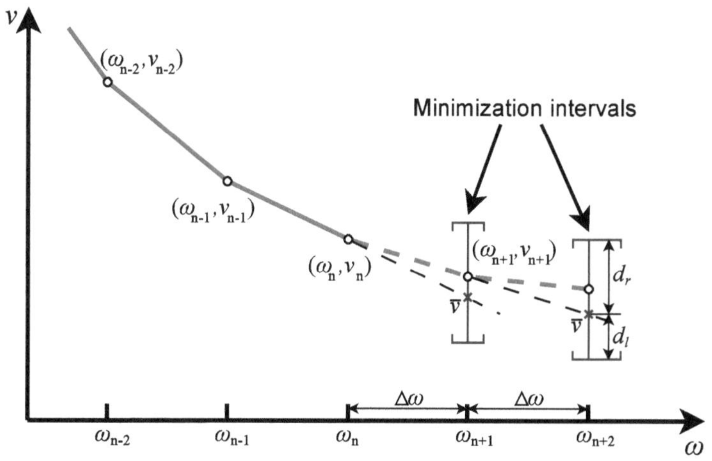

Figure 3.7: Extension of a dispersion curve to the interval $[\omega_n, \omega_{n+2}]$ with the step $\Delta\omega$.

if the interval is too large so that the fitting function has several minima on it. In this case the point may jump to another curve corresponding to another wave mode. In order to avoid it, it is suggested to take rather a small minimization interval (by taking smaller values for d_l and d_r) and to calculate with a smaller step in frequency. The computer program implementing the method allows user to look at the wave characteristics at every found point and thus to detect a jump to another wave mode if it takes place.

Remark. Instead of using a linear extrapolation to build the minimization interval we could use extrapolations of higher order. However, in practice there is no necessity of it, because the corrections of higher order extrapolations are insignificant in comparison with the interval length and the reliability of the method based on the linear extrapolation can always be improved by decreasing the frequency step $\Delta\omega$.

3.10 Computer Implementation

This section describes issues specific to the computer implementation of the model. The program is written in c++ The implementation of the graphical user interface is based on the `Qt` library, which makes the program portable to a wide range of platforms including Linux and Windows. The main numerical procedures exploit `LAPACK`.

3.10.1 Numerical Issues

The numerical implementation of the model is quite straightforward but requires significant efforts to ensure the numerical stability. Consider first the calculation of a fitting function at a given frequency. We sketch it here briefly, details are discussed in previous sections.

First, for every medium of the structure considered (layer or surrounding medium) a second order system of the linear differential equation for amplitude functions is constructed. In can be written in the general form as follows:

$$M\ddot{\boldsymbol{q}} + K\dot{\boldsymbol{q}} + L\boldsymbol{q} = 0. \tag{3.65}$$

The exact form of the matrices M, K and L depends on the material type (e.g., see (3.8), (3.31) and (3.44)). By inverting the matrix M and introducing the state vector

$$\boldsymbol{s} := \begin{pmatrix} \boldsymbol{q} \\ \dot{\boldsymbol{q}} \end{pmatrix},$$

the system above is reduced to the normal form:

$$\dot{\boldsymbol{s}} = A\boldsymbol{s}, \tag{3.66}$$

where

$$A := \begin{pmatrix} 0 & \mathbf{I} \\ M^{-1}L & M^{-1}K \end{pmatrix}.$$

Denote the dimension of A by m. m is either 12 or 16. Strictly speaking, it can also be 4 for a dielectric medium (see Section 3.5), but in that case the solution is specified explicitly and it is more efficiently to treat this case separately. For the exact form of A see e.g. (3.10), (3.32), and (3.45). The solution to (3.66) is expressed by the eigenvalues

$\{\lambda_j\}_{j=1}^m$ and eigenvectors $\{\boldsymbol{p}^j\}_{j=1}^m$ of A as follows (see for example (3.12)):

$$\begin{aligned}\boldsymbol{s}(x_3) = & \sum_{j=1}^{n} c_j\, \boldsymbol{p}^j e^{\lambda_j \kappa x_3} + \\ & + \sum_{j=\frac{n}{2}+1}^{m} \Big(c_{2j-1} \Big[\Re \boldsymbol{p}^{2j} \cos(\Re \lambda_{2j} \kappa x_3) - \Im \boldsymbol{p}^{2j} \sin(\Im \lambda_{2j} \kappa x_3)\Big] + \\ & \qquad + c_{2j} \Big[\Im \boldsymbol{p}^{2j} \cos(\Re \lambda_{2j} \kappa x_3) + \Re \boldsymbol{p}^{2j} \sin(\Im \lambda_{2j} \kappa x_3)\Big]\Big),\end{aligned} \tag{3.67}$$

This representation is then used to express the physical quantities involved into the boundary and interface conditions through the unknown coefficients $\{c_j\}_{j=1}^m$. Equating the appropriate quantities at the interfaces and free surfaces (if present) yields a linear system for unknown coefficients (see 3.20) in the form

$$F\boldsymbol{c} = 0,$$

where F is called the fitting matrix of the structure and $\boldsymbol{c}$ is the vector of the unknown coefficients collected from all the layers. By definition, the value of the fitting function is

$$\frac{\lambda_{min}(F^T F)}{\lambda_{max}(F^T F)}.$$

If this value is zero, the eigenvector $\boldsymbol{c}$ corresponding to the eigenvalue $\lambda_{min}(F^T F)$ contains the sought coefficients determining the wave (see Proposition 3.2). Summarizing, we can indicate the following three operations critical from the numerical point of view:

1. The step from (3.65) to (3.66). This is done for each layer and involves the inversion of the matrix M. This matrix is always invertible (and even positive-definite) as shown in the corresponding propositions above. Its dimension varies from 6×6 for conductive elastic, fluid, and bristle-like materials to 8×8 for piezoelectric materials.

2. The calculation of the eigenvalues and eigenvectors of A in (3.66). The dimension of the matrix is either 12×12 or 16×16 depending on the material.

3. The calculation of the fitting function on the base of the composed fitting matrix F. This involves the computation of the eigenvalues and (possibly) eigenvectors of $F^T F$. The dimension of the matrix $F^T F$ may be up to $16N \times 16N$, where N is the number of layers.

Computational Complexity. The three procedures above have the highest computational complexity. The first and the second ones require $N\mathcal{O}(16^3)$ operations. The third one is the most time-consuming, it requires $\mathcal{O}([16N]^3)$ operations. Hence the total computational complexity of the calculation of the fitting function at a single point is $\mathcal{O}([16N]^3)$. The symmetry of the matrix $F^T F$ allows to reduce the number of operations, but the order remains at $\mathcal{O}([16N]^3)$. Since the number of layers N lies in the range of 2 to 7 for most of the applications, the runtime needed for calculation of the fitting function at 1000 points does not usually exceed a few seconds. Thus the computational efficiency of the algorithm does not need any further improvement.

Round-off error. The main problem of the numerical implementation is the computational inaccuracy caused by the rounding error. The major source of this error are matrices containing elements of too different orders. The inversion or calculation of eigenvalues of such matrices are done with a significant inaccuracy. In the worst case the round-off error may completely kill the contribution of the small-order terms. The inaccurate results are then used in further calculation causing a serious discrepancy in the final outcome. To address this problem, we have used a number of tricks and scaling strategies as listed below:

- **Differentiation with respect to (κx_3).** The systems of differentiatonal equations for amplitudes (see (3.10), (3.32), (3.45)) are built with respect to the variable $\tilde{x_3} := \kappa x_3$ instead of x_3. This leads to a significantly more balanced matrices of the systems. For example, the matrix A defined by (3.11) describes the system (3.10) for the amplitudes $\boldsymbol{a}$ and $\boldsymbol{b}$ as functions of $\tilde{x_3}$. Now, suppose that the differentiation are done with respect to the variable x_3. Then the matrix of the system takes the form

$$\left(\begin{array}{cc|cc} 0 & 0 & \mathbf{I}_3 & 0 \\ 0 & 0 & 0 & \mathbf{I}_3 \\ \hline \kappa\, G_{\cdot 3\cdot 3}^{-1}\,(G_{\cdot 1\cdot 1} - \varrho v^2\,\mathbf{I}_3) & 0 & 0 & -\,\kappa^2\, G_{\cdot 3\cdot 3}^{-1}\,(G_{\cdot 1\cdot 3} + G_{\cdot 3\cdot 1}) \\ 0 & \kappa\, G_{\cdot 3\cdot 3}^{-1}\,(G_{\cdot 1\cdot 1} - \varrho v^2\,\mathbf{I}_3) & \kappa^2\, G_{\cdot 3\cdot 3}^{-1}\,(G_{\cdot 1\cdot 3} + G_{\cdot 3\cdot 1}) & 0 \end{array}\right)$$

  Though the amplitude functions found by this system are eventually the same from the mathematical point of view, the computed solution in the latter case are less accurate due to the factors κ ($\kappa \approx 10^5$) and κ^2 that introduce significant differences in orders of matrix entries.

- **Scaling the electric fields.** When dealing with piezoelectric and dielectric layers the matrices M, K, L are filled with combinations of components of the stiffness tensor G and the dielectric tensor ϵ (see (3.31)). The main components of G are usually of the order 10^{10}, the ones of ϵ are of the order 10^{-11}. To minimize this difference in magnitudes we substitute $c\tilde{\phi}$ for the electric potential ϕ and choose the coefficient c in such a way that the components of $c\epsilon$ are of the order 10^{10}. We reformulate then the system for the function $\tilde{\phi}$ in place of ϕ. The matrices of the obtained system are better balanced.

- **Scaling the eigenvectors for fluid.** The eigenvectors of the matrix A from (3.66) are initially normalized. This is however not optimal when treating the solid-fluid interface. Consider the contact conditions (3.48) and (3.49). When the eigenvectors for the contacting fluid and solid layers are of the norm 1 the contriubution in the fitting matrix of the right hand side is of the order 1, while the terms on the left-hand side are of the order $10^8 - 10^{10}$. To avoid this disbalance we multiply the normalized eigenvectors for the fluid by ω. This is exactly the factor that arises on the left-hand side in (3.48) after differentiation of the displacement $\boldsymbol{u}$ with respect to time.

- **Calculation in local coordinate system.** Recall that the physical quantities involved in interface and boundary conditions are expressed by using the representation (3.67). The terms at unknown coefficients c_j are calculated at interface planes and free surfaces and contribute then to the fitting matrix F. The origin of a numerical problem here are the exponential terms $\boldsymbol{p}^j e^{\lambda_j \kappa x_3}$ with $\lambda_j \in \mathbb{R}$. The ones calculated at higher interface planes are of higher order than those calculated for example at the interface plane $x_3 = 0$. This difference may become significant when many thick layers are present. This may lead to serious distortions in order of entries of the fitting matrix, thereby spoiling the computational accuracy.

  In order to avoid this situation we do all the calculations in a coordinate system local for each layer. This local coordinate system is built by shifting the origin to the bottom surface of the layer. When proceeding this way, the order of exponential terms that appear in the fitting matrix does not grow with the height of the interface plane. Note that the local calculation of physical quantities does not change their values and does not lead to a different result. Indeed, the exponential terms appearing in the fitting matrix become smaller, but the coefficients c_j at these terms scale correspondingly. Therefore, from the mathematical point of view, the result remains the same. But from the computational point of view the calculation in the local coordinate system allows to find it more accurate.

- **Exploiting singular value decomposition.** The fitting function is defined as the ratio of the minimal eigenvalue of F^TF to the maximal one. A straightforward way to find it would be to compute the product F^TF and then to find its eigenvalues. However, this approach is numerically unstable, especially for eigenvalues close to zero (see [30]). The mere multiplication of F^T by F doubles the difference in orders of entries. Therefore, instead of this, we find the eigenvalues of F^TF by exploiting the singular value decomposition of F. Note that F is a real matrix. Suppose it has dimensions $m \times n$. Its singular value decomposition is then of the form

$$F = U\Sigma V^T,$$

where $U \in \mathbb{R}^{m\times m}$ and $V \in \mathbb{R}^{n\times n}$ are orthogonal matrices and $\Sigma \in \mathbb{R}^{m\times n}$ has singular values of F on the diagonal and zeros off the diagonal. The eigenvalues of F^TF are then the squares of the singular values of F. Indeed, the eigenvalue decomposition of F^TF satisfies

$$F^TF = V\Sigma^TU^T\,U\Sigma V^T = V(\Sigma^T\Sigma)V^T.$$

This relation implies, moreover, that the columns of V are eigenvectors of F^TF. Denote the maximal and minimal singular values by σ_{max} and σ_{min} respectively. Then the value of the fitting function is

$$\frac{\lambda_{min}}{\lambda_{max}} = \left(\frac{\sigma_{min}}{\sigma_{max}}\right)^2.$$

Actually we could even define the fitting function as the ratio of σ_{min} to σ_{max} (without square). This ratio is non-dimensional as well and therefore has the same advantages. This choice is a matter of taste.

Note that the tricks above do not just improve the computational accuracy. Some of them are absoulutely crucial for obtaining any resonable result at all.

3.10.2 Program Description by Example

In this section we briefly describe the program implementing the model and demonstrate some features on an example of a real structure.

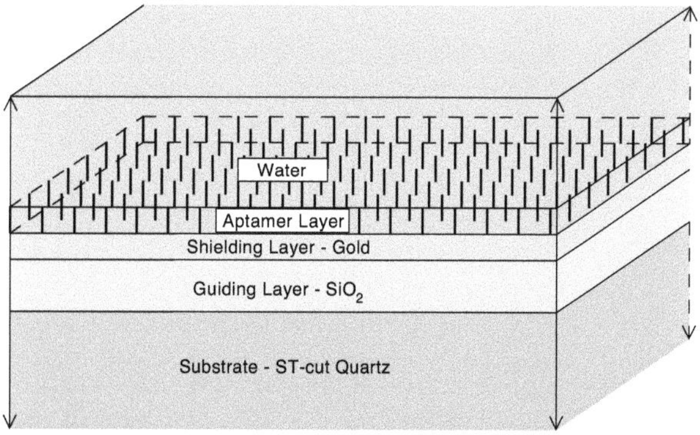

Figure 3.8: Example 1. A typical biosensor structure.

Suppose we would like to investigate Love acoustic waves in the structure depicted in Figure 3.8. This structure is one of the possible choices for the biosensor mentioned in Section 1.1. It consists of a thick piezoelectric substrate made of `ST-cut` of α-`quartz`, a guiding layer made of `SiO`$_2$, and a very thin gold shielding layer covered with bristle-like aptamer layer surrounded by water (for the explanation of aptamers, see Section 2.1). We demonstrate now the program on this example. The usage of the program includes the following steps:

1. Specification of the structure and wave frequency.
2. Calculation of the fitting function on an interval.
3. Localization of a root of the fitting function and determining it precisely. This yields the value of the phase velocity.
4. (Optional) Examination of the wave mode.
5. (Optional) Building the dispersion curve for the found wave mode.

1. Specification of the structure and wave frequency. The structure setup consists of the description of the ordered set of layers and the choice of the wave frequency. The layers are assumed to be ordered from the bottom towards the top so that each next layer lies on the previous one. Since the gravity force is not taken into account, this direction is

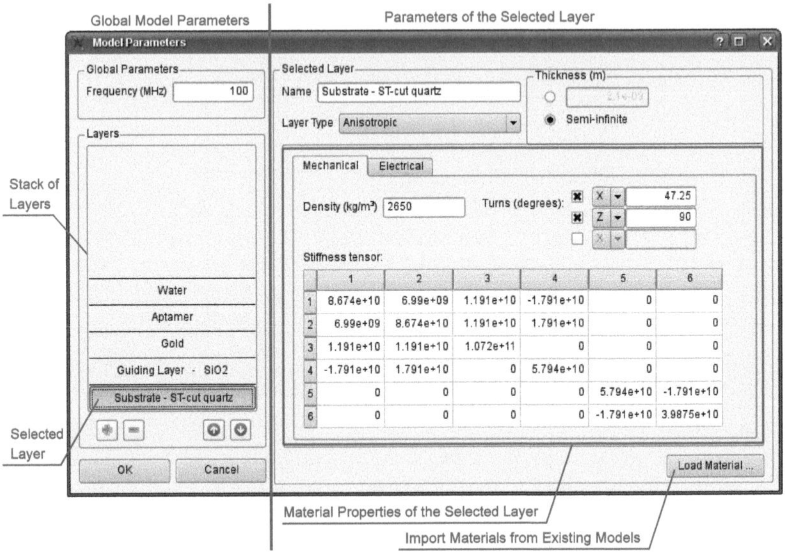

	1	2	3	4	5	6
1	8.674e+10	6.99e+09	1.191e+10	-1.791e+10	0	0
2	6.99e+09	8.674e+10	1.191e+10	1.791e+10	0	0
3	1.191e+10	1.191e+10	1.072e+11	0	0	0
4	-1.791e+10	1.791e+10	0	5.794e+10	0	0
5	0	0	0	0	5.794e+10	-1.791e+10
6	0	0	0	0	-1.791e+10	3.9875e+10

Figure 3.9: Dialog for setting model parameters. Parameters for the substrate.

relative. Therefore the reversed order of the layers should yield the same results. All the layers except for the bottom and the top ones must be of a finite thickness. The bottom and the top layers may be either of a finite thickness or occupy half spaces.

The parameters of the model are set in the dialog `"Model Parameters"` available through the menu item `Action -> Specify the Structure`. It is depicted in Figure 3.9. The dialog window consists of two parts. In the left part the set of layers and the frequency are specified. In the right part one can input or edit parameters for each layer. The content of the right part depends on the material type of the currently selected layer. Currently the following material types are supported:

- Isotropic (elastic/piezoelectric)
- Anisotropic (elastic/piezoelectric)
- Fluid
- Dielectric medium

- Bristle-like
- Periodic multilayers

The right part of the dialog in Figure 3.9 shows the parameters for the substrate. In our example, it is a semi-infinite piezoelectric layer. The stiffness, dielectric, and piezoelectric tensors are specified in the reference coordinate system of the crystal. The orientation of the material is described in terms of successive rotations of the reference system. In our example, the piezoelectric crystal is first rotated by 47.25 degrees around the X-axes and then by 90 degrees around the Z-axes. This corresponds to the so-called `ST-cut`. The order of rotations is important.

All the parameters are specified in SI units. The only exception is the frequency that is specified in MHz instead of Hz. All the tensors are specified in Voigt's notation. For the sake of convinience the program provides the feature to extract material parameters from existing saved models. This can spare a lot of time especially when dealing with piezoelectric materials.

The Figure 3.10 depicts the dialog with the settings for the aptamer layer. The parameters of bristle-like layers consist of the elastic parameters of the bristle, parameters of the surrounding fluid and the geometrical parameters of the elementary periodic cell containing a single bristle. Currently two shapes of bristles are supported - round and rectangular. The parameters `width` and `height` specifiy the relative size of a bristle. As the parameters of the structure are submitted, the cell equation is numerically solved and the parameters of the homogenized material are determined (see Section 3.7). The solving takes place just once for every set of parameters and the resulting tensors are then cashed for future use without recalculations. The cell equation is solved "on the fly" by the finite element method on a triangular grid automatically generated for a given cell.

2. Calculation of the fitting function. As the physical setup is done we can proceed with the calculation of the fitting function. To do this we specify the velocity interval in the dialog `Interval` called by choosing the menu item `Action -> Calculate`. The dialog takes three parameters: the starting velocity, the finishing velocity, and the number of divisions (see Figure 3.11). The approximate range of values for the sought wave velocity is usually known. In our case we take the interval $[3000, 5000]$ and calculate the fitting function with the step 2 m/s. Its graph is depicted in Figure 3.12.

3. Localization of a root and determining it precisely. Usually the interval for the preliminary calculation is quite large and hence the function has several roots on it.

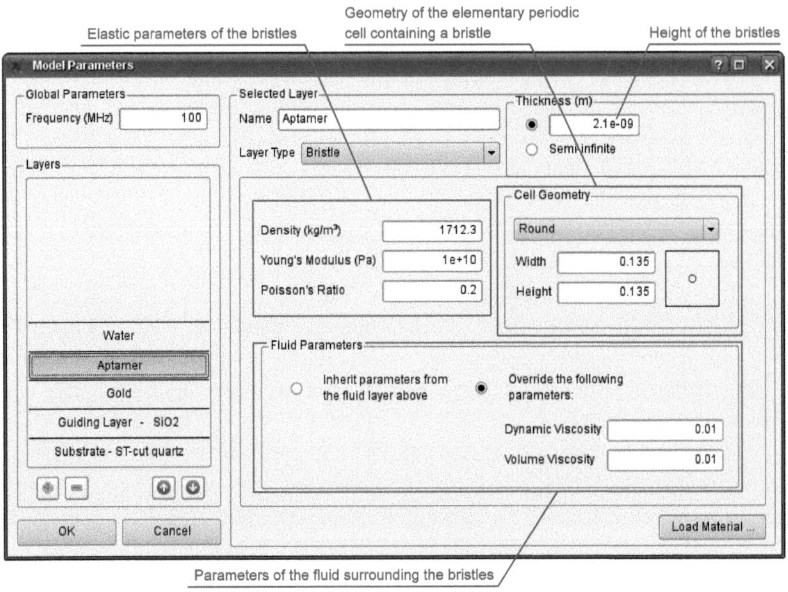

Figure 3.10: Dialog for setting model parameters. Parameters for the aptamers.

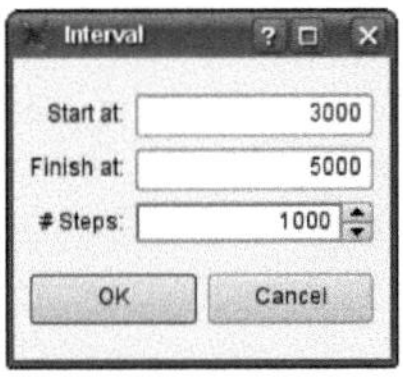

Figure 3.11: Dialog for setting the calculation interval.

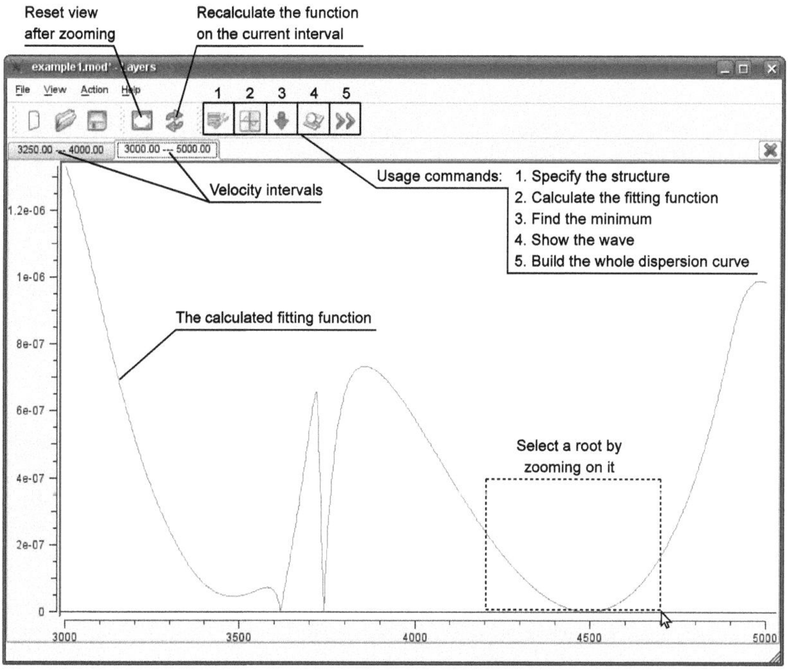

Figure 3.12: The main window of the program.

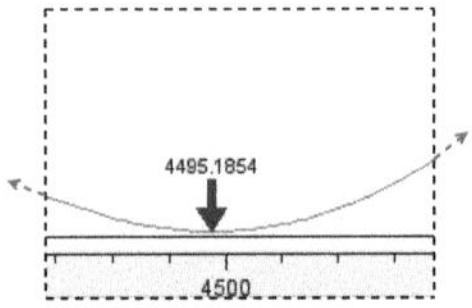

Figure 3.13: Local minimum of the fitting function.

Different roots correspond to different wave modes. To investigate a particular root we determine a subinterval containing it by zooming on the root (see Figure 3.12). In our case we select the root on the right, because it will give us a Love wave. One can check that the other two roots correspond to Rayleigh waves. When the root is localized it can be found precisely. This is done by the command `Find the minimum` started from the menu `Action` or from the toolbar. The result is shown in Figure 3.13. The phase velocity in our case is 4495.1854 m/s.

4. Examination of the wave mode. When the root of the fitting function is found, all the characteristics of the corresponding wave can easily be established. The dialog in Figure 3.14 shows the most important properties of the wave. Among other things it shows the displacements in layers depending on the phase. This enables to determine the wave mode. In our case only the u_2 component of the displacements is not zero, i.e. only shear displacements parallel to the interface planes are present. This corresponds to a Love wave. By the same way one can find out that the root of the calculated fitting funciton on the left yields a Rayleigh wave. Note that the actual thickness of the layers is not taken into account when drawing the displacements graphs. Otherwise very thin layers would not be visible at all. For the semi-infinite layers like substrate the graphs are drawn only up to depth of 5 wavelengths. This is usually more than enough because by construction the amplitudes decay exponentially with the depth.

5. Building the dispersion curve for the found wave. The found velocity in pair with the frequency yields a single point on the dispersion curve. We proceed then with the construction of the dispersion curve on some interval by the algorithm described in Section 3.9. We start with the point found at frequency 100 MHz and build the curve up to 150 MHz with the step 5 MHz. For the algorithm's parameters d_l and d_r responsible for the expectation velocity interval, we take 50 m/s. The dialog for the curve construction with the resulting curve is depicted in Figure 3.15. We can now continue to extend the function in any direction starting from any found point with an arbitrary step. For each

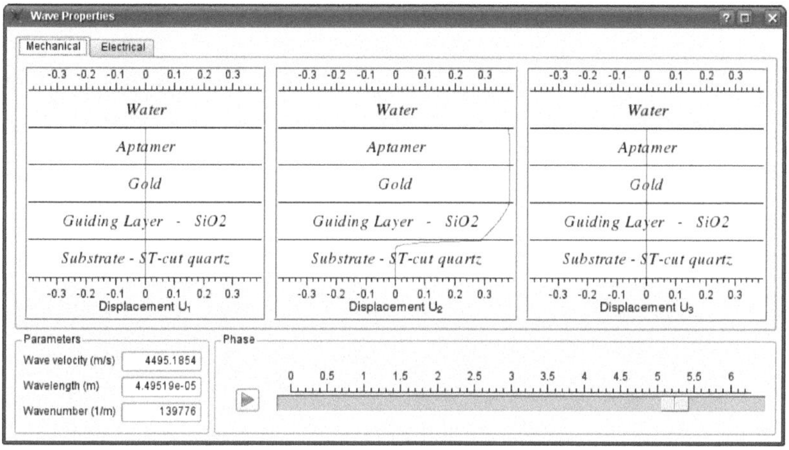

Figure 3.14: Properties of the found wave.

found point we can open the wave properties dialog and take a look at the corresponding wave. This way we can ensure that no jump to another wave mode occurred during the extension process. And if such a jump takes place the corresponding points can be deleted and the curve building can be repeated more carefully with a smaller frequency step. These features make a construction of dispersion curves simple, fast, and reliable.

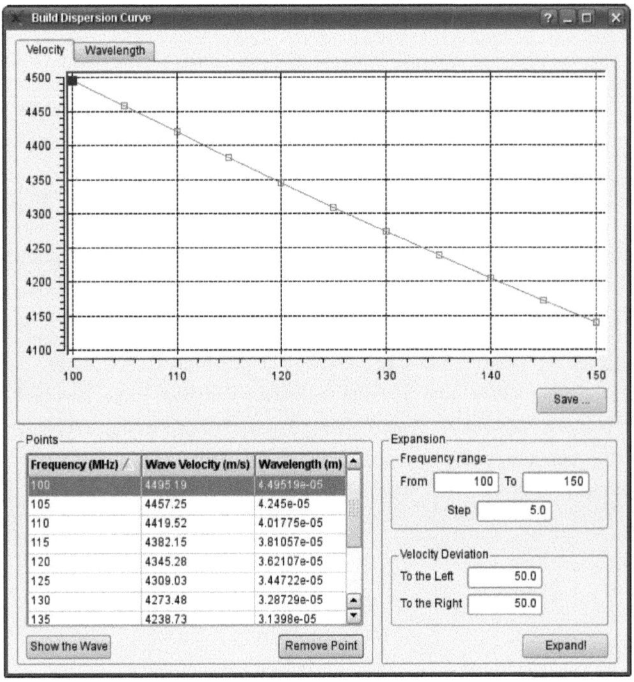

Figure 3.15: Dialog for building dispersion curve.

4 Homogenization of Linear Systems of Elasticity

4.1 Introduction

This chapter is devoted to the modeling of composite elastic periodic materials by exploiting the theory of homogenization. The materials under consideration are composed of several anisotropic materials mixed at the microscopic level. Besides, we assume that they have a periodic microstructure (see Figure 4.1) and the cell of periodicity is cubic. The smallness of the cell makes the direct modeling of the elastic behaviour of such materials impossible. For example, the cell might be smaller than the length of an acoustic wave. To overcome this difficulty one replaces the original composite material by the homogenized one that is described by the limiting equation obtained as the cell size ε goes to zero. The limiting equations for homogenized linear systems of elasticity are not new. They can be found for example in [56], though without derivation, and in [11], where they are obtained by Tartar's method of oscillating test functions (see [68, 69]). We derive these equations by the two-scale method of Ngutseng (see [54]) and Allaire (see [3]). The derivation follows [50], where the analogous results are obtained for the scalar case.

Our original motivation for the investigating homogenized elastic structures was the necessity of the modelling the wave propogation in so-called multilayers, i.e. in laminated materials with periodic microstructure (see Figure 4.2). Such materials can be considered as a special case of composite periodic materials, where the homogenization takes place only in one direction. This restriction causes significant simplifications in the cell equation and enables us to derive an explicit numerical scheme to calculating the effective material moduli of the homogenized material.

A natural questions that arises when replacing the original composed material with the homogenized one is how close is the solution of the homogenized problem to the solution of the original one. It is shown in [56] that if the right-hand side belongs to H^1 and the boundary between the composing materials is smooth, then the difference between the original solution and the first approximation estimated in H^1-norm is $\mathcal{O}(\sqrt{\varepsilon})$. However in real applications, the right-hand side is not weakly-differentiable as a rule. This motivated us to derive an estimate for the case when the right-hand side is in L^2. We also drop the assumption of smooth boundaries between the materials replacing it by a weaker one. These changes resulted in a weaker estimate making it $\mathcal{O}(\sqrt[3]{\varepsilon})$.

The chapter is organized as follows. In Section 4.3 we derive the limiting equations and investigate their properties. The case of laminated structures is considered in Section 4.4. In this section we establish an explicit formula for the numerical calculation of the stiffness tensor of the homogenized system. Finally, Section 4.5 is devoted to the investigation of the rate of convergence.

4.2 Notation

We use the following notation throughout the chapter:

$Y = [0,1]^3 \subset \mathbb{R}^3$	is a closed unit cube;
$\langle a \rangle_U = \dfrac{1}{\lvert U \rvert} \displaystyle\int_U a(\xi) d\xi$	is the mean value of a function a over a domain U;
$\mathcal{C}^n_\#(Y)$	is the space of all Y-periodic $\mathcal{C}^n(\mathbb{R}^3)$ functions;
$L^2_\#(Y)$	is the completion of $\mathcal{C}^\infty_\#(Y)$ with respect to the norm of $L^2(Y)$;
$H^1_\#(Y)$	is the completion of $\mathcal{C}^\infty_\#(Y)$ with respect to the norm of $H^1(Y)$ with the zero mean value, i.e. $\forall u \in H^1_\#(Y)\ \langle u \rangle = 0$;
$H(\mathrm{div};\Omega)$	$:= \left\{ f \in L^2(\Omega;\mathbb{R}^3) \,\middle|\, \mathrm{div} f \in L^2(\Omega) \right\}$;
δ_{ij}	is the Kronecker symbol.

If not otherwise stated, the indices i, j, k, l, m, n assume values 1, 2, 3. Throughout this chapter we adopt the Einstein summation convention, i.e. we sum over repeated indices.

For the sake of convenience we use C as a generic positive constant that can take different values at different occurrences.

Throughout the chapter we operate with sequences $\{\varepsilon_n\}_{n=1}^\infty$ of positive real numbers converging to zero as $n \to \infty$. Whenever there is no risk of ambiguity we omit the index n and write just ε. Moreover, in most of the cases we keep the same notation for the extracted subsequences.

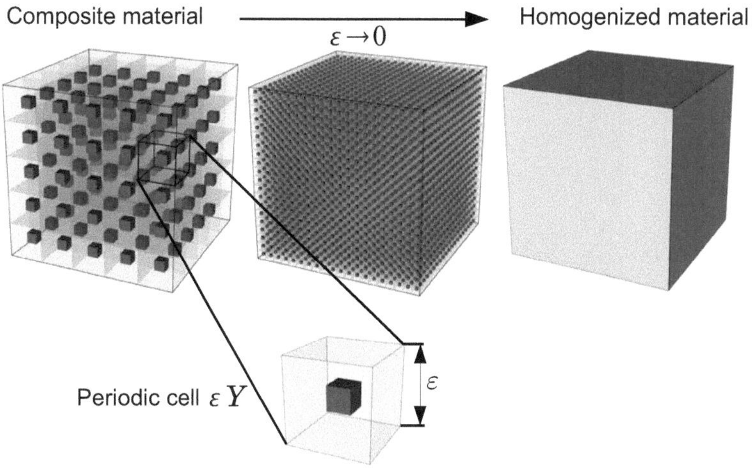

Figure 4.1: Homogenization approach.

4.3 Limiting Equations

Let Ω be an open bounded area in $\mathbb{R}^3$ with Lipschitz boundary $\partial\Omega$ and ε be the size of the cubic cell. The static linear elasticity equation augmented with the zero Dirichlet boundary condition reads:

$$\begin{cases} -\dfrac{\partial}{\partial x_l}\left(G^{\varepsilon}_{ijkl}(x)\dfrac{\partial u^{\varepsilon}_i}{\partial x_j}\right) = f_k(x) & \text{in } \Omega, \\ u^{\varepsilon} = 0 & \text{on } \partial\Omega. \end{cases} \tag{4.1}$$

Here $u^{\varepsilon}(x)$ is the displacement vector, G^{ε} is the elastic stiffness tensor. Since the material consists of periodically repeated cells composed of materials with constant properties, the tensor G^{ε} is a εY-periodic piecewise constant function of x. To make the material parameters independent of ε, we move this dependence to the argument by introducing

$$G_{ijkl}(x) := G^{\varepsilon}_{ijkl}(\varepsilon x).$$

Further, assume $f \in L^2(\Omega; \mathbb{R}^3)$. The weak formulation of (4.1) is then

$$\begin{cases} \int\limits_\Omega G_{ijkl}\left(\frac{x}{\varepsilon}\right) \frac{\partial u_i^\varepsilon}{\partial x_j}(x) \frac{\partial v_k}{\partial x_l}(x)\, dx = \int\limits_\Omega f_k(x) v_k(x) dx \qquad \forall v \in H_0^1(\Omega; \mathbb{R}^3), \\ u^\varepsilon \in H_0^1(\Omega; \mathbb{R}^3). \end{cases} \tag{4.2}$$

Let us define the bilinear form $a^\varepsilon : H^1(\Omega; \mathbb{R}^3) \times H^1(\Omega; \mathbb{R}^3) \mapsto \mathbb{R}$ by

$$a^\varepsilon(u, v) := \int\limits_\Omega G_{ijkl}\left(\frac{x}{\varepsilon}\right) \frac{\partial u_i}{\partial x_j}(x) \frac{\partial v_k}{\partial x_l}(x)\, dx. \tag{4.3}$$

The static elasticity problem for the composite material read then as follows.

Problem 4.1. *Find $u^\varepsilon \in H_0^1(\Omega; \mathbb{R}^3)$ such that*

$$a^\varepsilon(u^\varepsilon, v) = \langle f, v \rangle_{L^2(\Omega;\mathbb{R}^3)} \qquad \forall v \in H_0^1(\Omega; \mathbb{R}^3).$$

The new defined tensor $G_{ijkl}(y)$ is Y-periodic. Due to mechanical considerations, for all $y \in Y$ it is symmetric and positive definite, i.e.

$$G_{ijkl}(y) = G_{klij}(y) = G_{jikl}(y)$$

and

$$G_{ijkl}(y)\xi_{ij}\xi_{kl} \geqslant C(y)\xi_{ij}\xi_{ij}$$

for all second-rand symmetric tensors ξ. Moreover, since $G_{ijkl}(y)$ is piecewise constant on Y, it takes only a finite number of values, the constant on the right-hand side can be made independent on y by taking the maximum over Y. Hence, we have

$$G_{ijkl}(y)\xi_{ij}\xi_{kl} \geqslant C\xi_{ij}\xi_{ij} \qquad \forall\, y \in Y,\ \xi\text{-symmetric}. \tag{4.4}$$

The symmetry of G enables us to rewrite a^ε as follows:

$$a^\varepsilon(u, v) = \int\limits_\Omega G_{ijkl}\left(\frac{x}{\varepsilon}\right) \epsilon_{ij}(u)\epsilon_{kl}(v)\, dx, \tag{4.5}$$

where $\epsilon(u)$ is the symmetric part of ∇u, that is,

$$\epsilon_{ij}(u) := \frac{1}{2}\left(\frac{\partial u_i}{\partial x_j} + \frac{\partial u_j}{\partial x_i}\right).$$

Applying (4.4) and Korn's inequality (see Lemma 2.3) to (4.5), we obtain

$$a^\varepsilon(u,u) \geqslant C_1 \int\limits_\Omega |\epsilon(u)|^2\,dx \geqslant C_2 \|u\|^2_{H^1(\Omega;\mathbb{R}^3)} \qquad \forall\, u \in H^1_0(\Omega;\mathbb{R}^3),$$

where the constants $C_1, C_2 > 0$ do not depend on ε. This means that a^ε is elliptic on $H^1_0(\Omega;\mathbb{R}^3) \times H^1_0(\Omega;\mathbb{R}^3)$ and the constant of ellipticity is independent on ε. The boundedness of a^ε follows from the boundedness of $G(y)$ on Y and holds for all functions from $H^1(\Omega;\mathbb{R}^3)$. That is,

$$|a^\varepsilon(u,v)| \leqslant C\|u\|^2_{H^1(\Omega;\mathbb{R}^3)}\|v\|^2_{H^1(\Omega;\mathbb{R}^3)} \qquad \forall\, u, v \in H^1(\Omega;\mathbb{R}^3).$$

The well-posedness of Problem 4.1 follows then from the Lax-Milgram theorem.

Theorem 4.1. *For all $\varepsilon > 0$ Problem 4.1 has a unique solution u^ε and*

$$\|u^\varepsilon\|_{H^1(\Omega;\mathbb{R}^3)} \leqslant C\|f\|_{L^2(\Omega;\mathbb{R}^3)} \tag{4.6}$$

where the constant $C > 0$ on the right-hand side does not depend on ε.

We investigate now the behaviour of u^ε as ε goes to zero. We will need the following definition.

Definition. Let $\{\varepsilon_n\}_{n=1}^\infty$ be a sequence of positive real numbers (most of the time we will omit the subscript n) converging to 0. A sequence $\{u^\varepsilon\} \subset L_2(\Omega)$ is said *two-scale convergent* to a limit $u^0 \in L^2(\Omega \times Y)$ if

$$\lim_{\varepsilon\to 0} \int\limits_\Omega u^\varepsilon(x)\psi\left(x, \frac{x}{\varepsilon}\right) dx = \int\limits_\Omega \int\limits_Y u^0(x,y)\psi(x,y)\,dy\,dx \tag{4.7}$$

for all $\psi \in L^2(\Omega; \mathcal{C}_\#(Y))$. We denote two-scale convergence by

$$u^\varepsilon \underset{2}{\rightharpoonup} u^0.$$

A sequence of vector-valued functions $\{u^\varepsilon\} \subset L^2(\Omega;\mathbb{R}^N)$ is said *two-scale convergent* to $u^0 \in L_2(\Omega \times Y;\mathbb{R}^N)$ if u^ε_i two-scale converges to u^0_i for all $i = 1, .., N$.

The following three results are well-known and can be found for example in [50] (Theorems 3, 8, and 13).

Theorem 4.2. *Let* $g \in L^2(\Omega; \mathcal{C}_\#(Y))$. *Then* $g\left(x, \dfrac{x}{\varepsilon}\right)$ *is a measurable function on* Ω *and*

$$\left\| g\left(x, \frac{x}{\varepsilon}\right)\right\|_{L^2(\Omega)} \leqslant \|g(x,y)\|_{L^2(\Omega;\mathcal{C}_\#(Y))}.$$

Theorem 4.3. *Let* $\{u^\varepsilon\}$ *be a sequence in* $L^2(\Omega)$ *that two-scale converges to* $u^0 \in L^2(\Omega \times Y)$. *Then*

$$\lim_{\varepsilon \to 0} \int_\Omega u^\varepsilon(x)\psi\left(\frac{x}{\varepsilon}, x\right) dx = \int_\Omega \int_Y u^0(x,y)\psi(y,x)\, dy\, dx$$

for all $\psi \in L^2_\#(Y; \mathcal{C}(\overline{\Omega}))$.

Theorem 4.4. *Let* $\{u^\varepsilon\}$ *be a sequence in* $H^1_0(\Omega)$ *such that*

$$u^\varepsilon \rightharpoonup u^0 \qquad \text{in } H^1(\Omega).$$

Then $\{u^\varepsilon\}$ *two-scale converges to* u *and there exist a subsequence* ε' *and* $u^1 \in L^2(\Omega; H^1_\#(Y))$ *such that*

$$\nabla u^{\varepsilon'} \underset{2}{\rightarrow} \nabla u^0 + \nabla_y u^1.$$

We are ready now to derive the limiting equations.

Theorem 4.5. *Let* $\{u^\varepsilon\}$ *be a sequence of solutions of Problem 4.1,* $\varepsilon \to 0$. *Then there is a subsequence of* $\{\varepsilon_n\}_{n=1}^\infty$ *(that we still denote* ε*),* $u^0 \in H^1_0(\Omega; \mathbb{R}^3)$*, and* $u^1 \in \left[L^2(\Omega; H^1_\#(Y))\right]^3$ *such that*

$$\begin{aligned} u^\varepsilon &\rightharpoonup u^0 && \text{in } H^1(\Omega; \mathbb{R}^3), \\ \nabla u^\varepsilon &\underset{2}{\rightarrow} \nabla u^0 + \nabla_y u^1, \end{aligned}$$

and (u^0, u^1) *satisfies*

$$\int_\Omega \int_Y G_{ijkl}(y) \left[\frac{\partial u_i^0}{\partial x_j}(x) + \frac{\partial u_i^1}{\partial y_j}(x,y)\right] \left[\frac{\partial v_k^0}{\partial x_l}(x) + \frac{\partial v_k^1}{\partial y_l}(x,y)\right] dy\, dx = \int_\Omega f_k(x) v_k^0(x) dx \tag{4.8}$$

for all $(v_0, v_1) \in H^1_0(\Omega; \mathbb{R}^3) \times \left[L^2(\Omega; H^1_\#(Y))\right]^3$.

Proof. First note that Theorem 4.1 implies that $\{u^\varepsilon\}$ is bounded. Since $H^1(\Omega;\mathbb{R}^3)$ is a reflexive space, we can extract a subsequence (that we still denote ε) such that

$$u^\varepsilon \rightharpoonup u^0 \qquad \text{in } H^1(\Omega;\mathbb{R}^3) \tag{4.9}$$

for some u^0. We apply now Theorem 4.4 and discover that there exist a subsequence (still denoted by ε) and $u^1 \in \left[L^2(\Omega; H^1_\#(Y))\right]^3$ such that

$$\nabla u^\varepsilon \underset{2}{\rightarrow} \nabla u^0 + \nabla_y u^1. \tag{4.10}$$

Let us now show that the limiting functions u^0 and u^1 satisfy (4.8). Consider v of the form

$$v(x) = v^0(x) + \varepsilon v^1\left(x, \frac{x}{\varepsilon}\right),$$

where $v^0 \in C^1_0(\Omega;\mathbb{R}^3)$ and $v^1 \in \left[\mathcal{C}^\infty_0(\Omega;\mathcal{C}^1_\#(Y))\right]^3$. Obviously $v \in H^1_0(\Omega;\mathbb{R}^3)$. Substituting v as a test function into the integral identity in (4.2) yields

$$\int_\Omega G_{ijkl}\left(\frac{x}{\varepsilon}\right)\frac{\partial u_i^\varepsilon}{\partial x_j}(x)\frac{\partial v_k^0}{\partial x_l}(x)\,dx + \varepsilon\int_\Omega G_{ijkl}\left(\frac{x}{\varepsilon}\right)\frac{\partial u_i^\varepsilon}{\partial x_j}(x)\frac{\partial v_k^1}{\partial x_l}\left(x,\frac{x}{\varepsilon}\right)dx +$$
$$+\int_\Omega G_{ijkl}\left(\frac{x}{\varepsilon}\right)\frac{\partial u_i^\varepsilon}{\partial x_j}(x)\frac{\partial v_k^1}{\partial y_l}\left(x,\frac{x}{\varepsilon}\right)dx = \int_\Omega f_k(x)v_k^0(x)dx + \varepsilon\int_\Omega f_k(x)v_k^1\left(x,\frac{x}{\varepsilon}\right)dx.$$

To treat the second term on the right-hand side, we use the Cauchy-Schwarz inequality and Theorem 4.2. We get

$$\varepsilon\left|\int_\Omega f_k(x)v_k^1\left(x,\frac{x}{\varepsilon}\right)\right|^2 \leqslant \varepsilon\,\|f\|^2_{L^2(\Omega;\mathbb{R}^3)}\left\|v_k^1\left(x,\frac{x}{\varepsilon}\right)\right\|^2_{L^2(\Omega;\mathbb{R}^3)} \leqslant$$
$$\leqslant \varepsilon\,\|f\|^2_{L^2(\Omega;\mathbb{R}^3)}\left\|v_k^1(x,y)\right\|^2_{L^2(\Omega;C_\#(Y))} \to 0.$$

This means that the second term on the right-hand side goes to zero. In order to pass the left-hand side to the limit, we exploit (4.9) and (4.10) and apply Theorem 4.3. This yields

$$\int_\Omega\int_Y G_{ijkl}(y)\left[\frac{\partial u_i^0}{\partial x_j}(x) + \frac{\partial u_i^1}{\partial y_j}(x,y)\right]\frac{\partial v_k^0}{\partial x_l}(x)\,dy\,dx$$
$$+\int_\Omega\int_Y G_{ijkl}(y)\left[\frac{\partial u_i^0}{\partial x_j}(x) + \frac{\partial u_i^1}{\partial y_j}(x,y)\right]\frac{\partial v_k^1}{\partial y_l}(x,y)\,dy\,dx = \int_\Omega f_k(x)v_k^0(x)dx.$$

Due to the density of $C_0^1(\Omega)$ in $H_0^1(\Omega)$ and $C_0^\infty(\Omega; C_\#^1(Y))$ in $L^2(\Omega; H_\#^1(Y))$, this implies that (4.8) holds for all $(v_0, v_1) \in H_0^1(\Omega;\mathbb{R}^3) \times \left[L^2(\Omega; H_\#^1(Y))\right]^3$.

□

Note that the convergence result is proved only for some subsequence of ε. In order to prove it for the whole sequence, we show that (4.8) is uniquely solvable and hence the limiting functions u^0 and u^1 are the same for all the subsequences. We will need the following version of Korn's inequality for periodic functions

Lemma 4.6 (Korn's inequality for periodic functions). *For all $v \in H_\#^1(Y;\mathbb{R}^3)$ the following estimate holds:*

$$\|v\|_{H_\#^1(Y;\mathbb{R}^3)} \leqslant C\|\epsilon^y(v)\|_{L^2(\Omega;\mathbb{R}^{3\times 3})},$$

where the constant C does not depend on v; $\epsilon^y(v)$ is the symmetric part of $\nabla_y v$, that is,

$$\epsilon_{ij}^y(v) := \frac{1}{2}\left(\frac{\partial v_i}{\partial y_j} + \frac{\partial v_j}{\partial y_i}\right).$$

The proof can be found for example in [56] (Theorem 2.9).

Theorem 4.7. *The equation (4.8) is uniquely solvable in $H_0^1(\Omega;\mathbb{R}^3) \times \left[L^2(\Omega; H_\#^1(Y))\right]^3$.*

Proof. The proof is based on the Lax-Milgram theorem. Let us define the Hilbert space

$$\mathcal{V} := H_0^1(\Omega;\mathbb{R}^3) \oplus \left[L^2(\Omega; H_\#^1(Y))\right]^3$$

and the bilinear form

$$b(U,V) := \int_\Omega \int_Y G_{ijkl}(y)\left[\frac{\partial u_i^0}{\partial x_j}(x) + \frac{\partial u_i^1}{\partial y_j}(x,y)\right]\left[\frac{\partial v_k^0}{\partial x_l}(x) + \frac{\partial v_k^1}{\partial y_l}(x,y)\right] dy\, dx$$

for all $U = (u^0, u^1), V = (v^0, v^1) \in \mathcal{V}$. In order to prove the statement, it is enough to show that $b(\cdot,\cdot)$ is bounded and $\mathcal{V}$-elliptic. Let us first prove the $\mathcal{V}$-ellipticity. Let $U = (u^0, u^1)$ be an arbitrary element of $\mathcal{V}$. Put

$$\xi_{ij}(x,y) := \epsilon_{ij}(u^0) + \epsilon_{ij}^y(u^1) = \frac{1}{2}\left(\frac{\partial u_i^0}{\partial x_j}(x) + \frac{\partial u_j^0}{\partial x_i}(x)\right) + \frac{1}{2}\left(\frac{\partial u_i^1}{\partial y_j}(x,y) + \frac{\partial u_j^1}{\partial y_i}(x,y)\right),$$

where $\epsilon_{ij}^y(u^1)$ is the symmetric part of $\nabla_y u^1$. The symmetry and positive-definiteness of G

imply

$$b(U,U) = \int_\Omega \int_Y G_{ijkl}(y)\, \xi_{ij}(x,y)\, \xi_{kl}(x,y) dy\, dx \geqslant C \int_\Omega \int_Y \xi_{ij}(x,y) \xi_{ij}(x,y) dy\, dx =$$

$$= C \int_\Omega |\epsilon(u^0)|^2 \int_Y dy\, dx + C \int_\Omega \int_Y |\epsilon^y(u^1)|^2 dy\, dx + C \int_\Omega \epsilon_{ij}(u^0) \int_Y \epsilon^y_{ij}(u^1) dy\, dx =$$
$$= C \|\epsilon(u^0)\|^2_{L^2(\Omega;\mathbb{R}^{3\times 3})} + C \int_\Omega \|\epsilon(u^1)\|^2_{L^2(Y;\mathbb{R}^{3\times 3})} dx.$$

The third term disappeared because

$$\int_Y \frac{\partial u_i^1}{\partial y_j}(x,y) dy = 0 \qquad \forall i,j \quad \forall x \in \Omega.$$

This can easily be proved by integrating by parts and using the Y-periodicity of u^1. The rest two terms are estimated by Korn's inequalities, that is, by Lemma 2.3 and Lemma 4.6. Hence we obtain

$$b(U,U) \geqslant C \left(\left\| u^0 \right\|^2_{H^1(\Omega;\mathbb{R}^3)} + \left\| u^1 \right\|^2_{\left[L^2(\Omega;H^1_\#(Y))\right]^3)} \right) = C \left\| (u^0, u^1) \right\|^2_{\mathcal{V}}.$$

The $\mathcal{V}$-ellipticity of $b(\cdot,\cdot)$ is thus established. The boundedness follows from the boundedness of G and is proved by the Cauchy-Schwarz inequality. We have

$$|b(U,V)| \leqslant \int_\Omega \int_Y |G_{ijkl}(y)| \left| \frac{\partial u_i^0}{\partial x_j}(x) + \frac{\partial u_i^1}{\partial y_j}(x,y) \right| \left| \frac{\partial v_k^0}{\partial x_l}(x) + \frac{\partial v_k^1}{\partial y_l}(x,y) \right| dy\, dx \leqslant$$

$$\leqslant C \left(\int_\Omega \int_Y \left| \nabla u^0 + \nabla_y u^1 \right|^2 dydx \right)^{1/2} \left(\int_\Omega \int_Y \left| \nabla v^0 + \nabla_y v^1 \right|^2 dydx \right)^{1/2} \leqslant$$

$$\leqslant 2C \left(\int_\Omega \left| \nabla u^0 \right|^2 dx + \int_\Omega \int_Y \left| \nabla_y u^1 \right|^2 dydx \right)^{1/2} \left(\int_\Omega \left| \nabla v^0 \right|^2 dx + \int_\Omega \int_Y \left| \nabla_y v^1 \right|^2 dydx \right)^{1/2} \leqslant$$

$$\leqslant C \|U\|_{\mathcal{V}} \|V\|_{\mathcal{V}}.$$

Hence, $b(\cdot,\cdot)$ is $\mathcal{V}$-elliptic and bounded. Applying the Lax-Milgram theorem, we see that (4.8) possesses a unique solution. □

Thus, the limiting functions are uniquely determined. Therefore the convergence takes place for the whole sequence of solutions. The following theorem provides a more convenient way to the calculation of the limiting functions u^0 and u^1.

Theorem 4.8. *Let $\{u^\varepsilon\}$ be a sequence of solutions of Problem 4.1, $u^0 \in H_0^1(\Omega;\mathbb{R}^3)$ and $u^1 \in \left[L^2(\Omega; H_\#^1(Y))\right]^3$ be the limiting functions, that is,*

$$\begin{aligned} u^\varepsilon &\rightharpoonup u^0 \qquad\qquad \textit{in } H^1(\Omega;\mathbb{R}^3),\\ \nabla u^\varepsilon &\underset{2}{\rightharpoonup} \nabla u^0 + \nabla_y u^1. \end{aligned}$$

Then u^0 and u^1 possess the following properties:

- *u^1 admits the representation*

$$u_i^1(x,y) = N_{imn}(y)\frac{\partial u_m^0}{\partial x_n}(x), \tag{4.11}$$

where $N_{\cdot mn}(y) \in H_\#^1(Y;\mathbb{R}^3)$ are unique solutions of the so-called cell equations

$$\int_Y G_{ijkl}(y)\frac{\partial N_{imn}}{\partial y_j}(y)\frac{\partial v_k^1}{\partial y_l}(y)\,dy = -\int_Y G_{mnkl}(y)\frac{\partial v_k^1}{\partial y_l}(y)\,dy \quad \forall v^1 \in H_\#^1(Y;\mathbb{R}^3). \tag{4.12}$$

- *u^0 satisfies the limiting equation*

$$\int_\Omega G_{mnkl}^{\mathbf{hom}}\frac{\partial u_m^0}{\partial x_n}(x)\frac{\partial v_k^0}{\partial x_l}(x)\,dx = \int_\Omega f_k v_k^0\,dx \qquad \forall v^0 \in H_0^1(\Omega;\mathbb{R}^3) \tag{4.13}$$

with

$$G_{mnkl}^{\mathbf{hom}} := \int_Y G_{mnkl}(y)dy + \int_Y G_{ijkl}(y)\frac{\partial N_{imn}}{\partial y_j}(y)dy. \tag{4.14}$$

Proof. As established in Theorem 4.5, u^0 and u^1 satisfy (4.8). Since v^0 and v_1 are independent, (4.8) splits up in two equations as follows:

$$\begin{aligned} &\int_\Omega\int_Y G_{ijkl}(y)\left[\frac{\partial u_i^0}{\partial x_j}(x) + \frac{\partial u_i^1}{\partial y_j}(x,y)\right]\frac{\partial v_k^0}{\partial x_l}(x)dy\,dx = \int_\Omega f_k(x)v_k^0(x)dx,\\ &\int_\Omega\int_Y G_{ijkl}(y)\left[\frac{\partial u_i^0}{\partial x_j}(x) + \frac{\partial u_i^1}{\partial y_j}(x,y)\right]\frac{\partial v_k^1}{\partial y_l}(x,y)dy\,dx = 0. \end{aligned} \tag{4.15}$$

We are looking for u_1 in the form

$$u^1(x,y) = N_{\cdot mn}(y)\frac{\partial u_m^0}{\partial x_n}(x), \tag{4.16}$$

with some unknown $N_{\cdot mn}(y) \in H^1(\Omega;\mathbb{R}^3)$. Substituting it in (4.15) yields

$$\int_\Omega\int_Y G_{ijkl}(y)\left[\delta_{mi}\delta_{nj} + \frac{\partial N_{imn}}{\partial y_j}(y)\right]\frac{\partial u_m^0}{\partial x_n}(x)\frac{\partial v_k^0}{\partial x_l}(x)dy\,dx = \int_\Omega f_k(x)v_k^0(x)dx,$$
$$\int_\Omega\int_Y G_{ijkl}(y)\left[\delta_{mi}\delta_{nj} + \frac{\partial N_{imn}}{\partial y_j}(y)\right]\frac{\partial u_m^0}{\partial x_n}(x)\frac{\partial v_k^1}{\partial y_l}(x,y)dy\,dx = 0.$$

Let us define

$$G_{mnkl}^{\mathtt{hom}} := \int_Y G_{ijkl}(y)\left[\delta_{im}\delta_{jn} + \frac{\partial N_{imn}}{\partial y_j}(y)\right]dy.$$

Note that this definition is equivalent to (4.14). Then the first equation takes the form

$$\int_\Omega\int_Y G_{mnkl}^{\mathtt{hom}}\frac{\partial u_m^0}{\partial x_n}(x)\frac{\partial v_k^0}{\partial x_l}(x)dy\,dx = \int_\Omega f_k(x)v_k^0(x)dx.$$

The second equation holds for all $v^1 \in \left[L^2(\Omega;H_\#^1(Y))\right]^3$, in particular for all v^1 that are constant with respect to x. In this case it can be rewritten as follows:

$$\int_\Omega \frac{\partial u_m^0}{\partial x_n}(x)\int_Y G_{ijkl}(y)\left[\delta_{mi}\delta_{nj} + \frac{\partial N_{imn}}{\partial y_j}(y)\right]\frac{\partial v_k^1}{\partial y_l}(y)dy\,dx = 0 \qquad \forall v^1 \in H_\#^1(Y;\mathbb{R}^3).$$

We require now that for all fixed m,n

$$\int_Y G_{ijkl}(y)\left[\delta_{mi}\delta_{nj} + \frac{\partial N_{imn}}{\partial y_j}(y)\right]\frac{\partial v_k^1}{\partial y_l}(y)dy = 0 \qquad \forall v^1 \in H_\#^1(Y;\mathbb{R}^3). \tag{4.17}$$

This is equivalent to (4.12), that is,

$$\int_Y G_{ijkl}(y)\frac{\partial N_{imn}}{\partial y_j}(y)\frac{\partial v_k^1}{\partial y_l}(y)dy = -\int_Y G_{mnkl}(y)\frac{\partial v_k^1}{\partial y_l}(y)dy \qquad \forall v^1 \in H_\#^1(Y;\mathbb{R}^3).$$

It is left to show that (4.12) is uniquely solvable in $H_\#^1(Y;\mathbb{R}^3)$ for all m,n. This is done again by the Lax-Milgram theorem. The proof of the ellipticity is similar to that of Theorem 4.7.

It is based on Lemma 4.6. The boundedness is obtained straightforwardly by applying the Cauchy-Schwarz inequality.

Thus, the existence and uniqueness of $N_{\cdot mn}$ are established, the suppositions (4.16) and (4.17) are justified.

□

As one can see the limiting equation (4.13) is of the same form as the elasticity equation for some constant material. We ensure now that the homogenized tensor G^{hom} possesses the same properties as a normal stiffness tensor.

Theorem 4.9. *Tensor G^{hom} is*

(i) symmetric, i.e.

$$G^{\mathrm{hom}}_{mnkl} = G^{\mathrm{hom}}_{nmkl} = G^{\mathrm{hom}}_{klmn}$$

(ii) positive-definite, i.e.

$$G^{\mathrm{hom}}_{mnkl}\xi_{mn}\xi_{kl} > 0$$

for all symmetric non-zero second-rang tensors ξ.

Proof. (i). The first equality follows from the corresponding symmetry of G. Since

$$G_{mnkl} = G_{nmkl}$$

and $N_{\cdot mn}$ are uniquely determined by (4.12), we have

$$N_{mn} = N_{nm}.$$

The equality $G^{\mathrm{hom}}_{mnkl} = G^{\mathrm{hom}}_{nmkl}$ follows then from the definition of G^{hom} (see (4.14)).

We prove now that $G^{\mathrm{hom}}_{mnkl} = G^{\mathrm{hom}}_{klmn}$. Since

$$G_{mnkl} = G_{klmn},$$

the first terms of G^{hom}_{mnkl} and G^{hom}_{klmn} in (4.14) coincide and it is enough to show that for all m, n, k, and l

$$\int_Y G_{ijkl}(y)\frac{\partial N_{imn}}{\partial y_j}(y)dy = \int_Y G_{pqmn}(y)\frac{\partial N_{pkl}}{\partial y_q}(y)dy, \tag{4.18}$$

where p and q are integer indices taking values from 1 to 3. We used them here instead of i and j to avoid the confusion in the following derivation. Let m, n, k, and l be fixed. Taking in (4.12) first $v^1 = N_{\cdot kl}$ and then $v^1 = N_{\cdot mn}$ and using the symmetry of G, we obtain

$$-\int_Y G_{pqmn}(y)\frac{\partial N_{pkl}}{\partial y_q}(y)dy = -\int_Y G_{mnpq}(y)\frac{\partial N_{pkl}}{\partial y_q}(y)dy =$$

$$= \int_Y G_{ijpq}(y)\frac{\partial N_{imn}}{\partial y_j}(y)\frac{\partial N_{pkl}}{\partial y_q}(y)\,dy = \int_Y G_{pqij}(y)\frac{\partial N_{pkl}}{\partial y_q}(y)\frac{\partial N_{imn}}{\partial y_j}(y)\,dy =$$

$$= -\int_Y G_{klij}(y)\frac{\partial N_{imn}}{\partial y_j}(y)dy = -\int_Y G_{ijkl}(y)\frac{\partial N_{imn}}{\partial y_j}(y)dy.$$

The equality (4.18) is thus proved and *(i)* is established.

(ii). Let ξ be a symmetric second-rang tensor. We have

$$G^{\mathtt{hom}}_{mnkl}\xi_{mn}\xi_{kl} = \xi_{mn}\int_Y G_{ijkl}(y)\left[\delta_{im}\delta_{jn} + \frac{\partial N_{imn}}{\partial y_j}(y)\right]\xi_{kl}\,dy. \tag{4.19}$$

Taking $v^1(y) = \xi_{pq}N_{\cdot pq}(y)$ in the cell equation in form (4.17) yields

$$\int_Y G_{ijkl}(y)\left[\delta_{mi}\delta_{nj} + \frac{\partial N_{imn}}{\partial y_j}(y)\right]\xi_{pq}\frac{\partial N_{kpq}}{\partial y_l}(y)dy = 0 \qquad \forall m,n.$$

Multiplying it by ξ_{mn} and adding to (4.19), we obtain

$$G^{\mathtt{hom}}_{mnkl}\xi_{mn}\xi_{kl} = \xi_{mn}\int_Y G_{ijkl}(y)\left[\delta_{im}\delta_{jn} + \frac{\partial N_{imn}}{\partial y_j}(y)\right]\left[\xi_{kl} + \xi_{pq}\frac{\partial N_{kpq}}{\partial y_l}(y)\right]dy -$$

$$= \int_Y G_{ijkl}(y)\left[\xi_{ij} + \xi_{mn}\frac{\partial N_{imn}}{\partial y_j}(y)\right]\left[\xi_{kl} + \xi_{pq}\frac{\partial N_{kpq}}{\partial y_l}(y)\right]dy = \int_Y G_{ijkl}(y)Z_{ij}(y)Z_{kl}(y)\,dy \geqslant 0,$$

where

$$Z_{ij}(y) := \xi_{ij} + \xi_{mn}\frac{\partial N_{imn}}{\partial y_j}(y). \tag{4.20}$$

The inequality is not strict because Z is not necessarily symmetric. Thus $G^{\mathtt{hom}}$ is non-negative. Let us show that it is strictly positive. Suppose

$$G^{\mathtt{hom}}_{mnkl}\xi_{mn}\xi_{kl} = 0.$$

This implies that

$$G_{ijkl}(y)Z_{ij}(y)Z_{kl}(y) = 0 \quad \text{a.e. in } Y.$$

Figure 4.2: Laminated periodic material.

Due to the symmetry of G this is equivalent to

$$G_{ijkl}(y)\frac{1}{2}\left(Z_{ij}(y)+Z_{ji}(y)\right)\frac{1}{2}\left(Z_{kl}(y)+Z_{lk}(y)\right)=0 \quad \text{a.e. in } Y.$$

Since G is positive definite for all $y \in Y$, this means that

$$\forall\, i,j \qquad Z_{ij}(y)+Z_{ji}(y)=0 \quad \text{a.e. in } Y.$$

Integrating this equality over Y and taking into account that N_{imn} are Y-periodic functions, we obtain

$$\forall\, i,j \qquad 0=\int\limits_Y\left(\xi_{ij}+\xi_{mn}\frac{\partial N_{imn}}{\partial y_j}(y)\right)dy+\int\limits_Y\left(\xi_{ji}+\xi_{mn}\frac{\partial N_{jmn}}{\partial y_i}(y)\right)dy=\xi_{ij}+\xi_{ji}=2\xi_{ij}.$$

Thus $\xi=0$. Therefore G^{hom} is positive definite.

□

4.4 Homogenization of Laminated Structures

In this section we consider composite materials periodic only in one direction and homogeneous in others (see Figure 4.2). We assume that the material is rotated in such a way that the periodic direction is parallel to x_3. This is a special case of periodic composite materials considered in the previous section. In this case the cell stiffness tensor G depends only on y_3 and independent on y_1 and y_2. This simplification enables us to derive explicit formulas for calculating G^{hom}.

We will need the following well-known result.

Lemma 4.10 (Poincaré's inequality). *Let U be a bounded connected domain in $\mathbb{R}^n$ having Lipschitz boundary, $1 \leqslant p \leqslant \infty$. Then there exists a constant C, depending only on n, p, and U, such that*

$$\|u - \langle u\rangle_U\|_{L^p(U)} \leqslant C\,\|\nabla u\|_{L^p(U)}$$

for each function $u \in W^{1,p}(U)$.

Theorem 4.11. *Suppose that the cell stiffness tensor $G(y)$ is independent of y_1 and y_2. Then for all m, n functions $N_{\cdot mn}$ are independent of y_1 and y_2 and satisfy the following one-dimensional cell equations:*

$$\int_0^1 G_{i3k3}(y_3)\frac{dN_{imn}}{dy_3}(y_3)\frac{dw_k}{dy_3}(y_3)\,dy_3 = -\int_0^1 G_{mnk3}(y_3)\frac{dw_k}{dy_3}(y_3)\,dy_3 \qquad \forall w \in H^1_{\#}([0,1];\mathbb{R}^3). \tag{4.21}$$

In this case, the homogenized stiffness tensor takes the form

$$G^{\mathrm{hom}}_{mnkl} = \int_0^1 G_{mnkl}(y_3)\,dy_3 + \int_0^1 G_{i3kl}(y_3)\frac{\partial N_{imn}}{\partial y_3}(y_3)\,dy_3. \tag{4.22}$$

Proof. Let us first show that (4.21) is uniquely solvable for all m, n. This can easily be done by the Lax-Milgram theorem. The boundedness of the corresponding bilinear form is established by the Cauchy-Schwarz inequality. The ellipticity is due to the positiveness of the matrix $G_{\cdot 3\cdot 3}$ and the Poincaré inequality (see Lemma 4.10). The positiveness of the matrix $G_{\cdot 3\cdot 3}$ follows from the positiveness of G. Indeed, for any vector ξ we have

$$G_{i3k3}\xi_i\xi_k = G_{ijkl}(\delta_{j3}\xi_i)(\delta_{l3}\xi_k) = G_{ijkl}Z_{ij}Z_{kl} \geqslant 0,$$

where $Z_{ij} := \delta_{j3}\xi_i$. The equality is reached only if Z is antisymmetric, that is,

$$\delta_{j3}\xi_i = -\delta_{i3}\xi_j.$$

In this case $\xi_1 = \delta_{33}\xi_1 = -\delta_{13}\xi_3 = 0$. Similarly $\xi_2 = 0$ and $\xi_3 = -\xi_3$, i.e. $\xi_3 = 0$. We discover therefore that $G_{i3k3}\xi_i\xi_k = 0$ implies $\xi = 0$. Thus the matrix $G_{\cdot 3\cdot 3}$ is positive definite and therefore by the Lax-Milgram theorem (4.21) is uniquely solvable.

Recall that $N_{\cdot mn}(y)$ are unique solutions of the general cell equation (4.12). If we show that the solutions of (4.21), considered as functions on Y, solve (4.12), it would mean that $N_{\cdot mn}$ are independent on y_1 and y_2 and can be found by (4.21). Let us substitute $N_{\cdot mn}(y_3)$ into (4.12) and check if the equality holds for any arbitrary $v^1 \in H^1_{\#}(Y;\mathbb{R}^3)$. We must check the following relation:

$$\int_Y G_{i3kl}(y_3)\frac{\partial N_{imn}}{\partial y_3}(y_3)\frac{\partial v^1_k}{\partial y_l}(y)\,dy \stackrel{?}{=} -\int_Y G_{mnkl}(y_3)\frac{\partial v^1_k}{\partial y_l}(y)\,dy.$$

This is equivalent to

$$\int_0^1 G_{i3kl}(y_3)\frac{\partial N_{imn}}{\partial y_3}(y_3)\left(\int_{[0,1]^2}\frac{\partial v^1_k}{\partial y_l}(y)\,dy_1dy_2\right)dy_3 \stackrel{?}{=} -\int_0^1 G_{mnkl}(y_3)\left(\int_{[0,1]^2}\frac{\partial v^1_k}{\partial y_l}(y)\,dy_1dy_2\right)dy_3.$$

By introducing

$$w(y_3) := \int_{[0,1]^2} v^1(y)\,dy_1dy_2 \quad \in H^1_{\#}([0,1];\mathbb{R}^3)$$

the relation can be rewritten as

$$\int_0^1 G_{i3k3}(y_3)\frac{\partial N_{imn}}{\partial y_3}(y_3)\frac{\partial w_k}{\partial y_3}(y_3)\,dy_3 \stackrel{?}{=} -\int_0^1 G_{mnk3}(y_3)\frac{\partial w_k}{\partial y_3}(y_3)\,dy_3.$$

This is exactly (4.21). Therefore the equality takes place. Thus $N_{\cdot mn}(y_3)$ are solutions of (4.12) and the theorem is proved.

□

Lemma 4.12. *Let $U := [0,1]$ and $a \in L^2_{\#}(U)$ be such that*

$$\int_U a(\xi)\frac{d\varphi}{d\xi}(\xi)\,d\xi = 0 \qquad \forall\varphi \in H^1_{\#}(U). \tag{4.23}$$

Then

$$a = \langle a\rangle_U \quad \text{in } L^2_{\#}(U),$$

that is, a is constant almost everywhere on U.

Proof. For all $\varphi \in H^1_\#(U)$ we have

$$\int_U \langle a\rangle_U \frac{d\varphi}{d\xi}(\xi)\,d\xi = \langle a\rangle_U \int_U \frac{d\phi}{d\xi}(\xi)\,d\xi = \langle a\rangle_U \cdot 0 = 0.$$

Subtracting this from (4.23), we obtain

$$\int_U [a(\xi) - \langle a\rangle_U]\,\frac{d\varphi}{d\xi}(\xi)\,d\xi = 0 \qquad \forall \varphi \in H^1_\#(U). \tag{4.24}$$

Let us take the function φ in the form

$$\varphi(\xi) = \int_0^\xi a(z)\,dz - \langle a\rangle_U \cdot \xi.$$

We claim that this function is 1-periodic. Indeed,

$$\begin{aligned}\varphi(\xi+1) &= \int_0^{\xi+1} a(z)\,dz - \langle a\rangle_U(\xi+1) = \varphi(\xi) + \int_\xi^{\xi+1} a(z)dz - \langle a\rangle_U = \\ &= \varphi(\xi) + \int_U a(z)dz - \langle a\rangle_U = \varphi(\xi).\end{aligned}$$

It can be directly proved that φ is weakly differentiable and

$$\frac{d\varphi}{d\xi}(\xi) = a(\xi) - \langle a\rangle_U \in L^2_\#(U).$$

We can then substitute it in (4.24). This yields

$$\int_U [a(\xi) - \langle a\rangle_U]\,[a(\xi) - \langle a\rangle_U]\,d\xi = 0,$$

which implies

$$a = \langle a\rangle_U \quad \text{in } L^2_\#(U).$$

□

We derive now an explicit formula for the calculation of $G^{\mathtt{hom}}$. First let us rewrite (4.21) as follows

$$\int_0^1 G_{ijk3}(y_3)\left[\delta_{mi}\delta_{nj}+\delta_{j3}\frac{dN_{imn}}{dy_3}(y_3)\right]\frac{dw_k}{dy_3}(y_3)\,dy_3=0.$$

Since this equality holds for all $w \in H^1_\#([0,1];\mathbb{R}^3)$, we can apply Lemma 4.12. We have then almost everywhere on $[0,1]$

$$G_{ijk3}(y_3)\left[\delta_{mi}\delta_{nj}+\delta_{j3}\frac{dN_{imn}}{dy_3}(y_3)\right]=D_{mnk} \qquad \forall m,n,k,$$

where D_{mnk} are some unknown constants. Simplifying the relation above, we obtain the following system of linear algebraic equations for every pair (m,n):

$$G_{i3k3}(y_3)\frac{dN_{imn}}{dy_3}(y_3)=D_{mnk}-G_{mnk3}(y_3).$$

Note that the matrix of this system $G_{i3k3}(y_3)$ is the same for all m and n. Moreover, it is positive definite (see the proof of Theorem 4.11) and therefore invertible. Hence it holds:

$$\frac{dN_{imn}}{dy_3}(y_3)=(G_{\cdot 3\cdot 3}(y_3))^{-1}_{ik}\left[D_{mnk}-G_{mnk3}(y_3)\right]. \tag{4.25}$$

We integrate now this expression over $[0,1]$. The left-hand side disappears because N_{imn} are 1-periodical. Then for the vectors $D_{mn\cdot}$ we obtain

$$\left(\int_0^1 (G_{\cdot 3\cdot 3}(y_3))^{-1}dy_3\right)D_{mn\cdot}=\int_Y (G_{\cdot 3\cdot 3}(y_3))^{-1}G_{mn\cdot 3}(y_3)dy_3.$$

Since $G_{\cdot 3\cdot 3}(y_3)$ is a symmetric positive definite matrix for all y_3, so is the inverse matrix $(G_{\cdot 3\cdot 3}(y_3))^{-1}$ and hence the matrix

$$\left(\int_0^1 (G_{\cdot 3\cdot 3}(y_3))^{-1}dy_3\right)$$

is also positive definite. We can then invert it and obtain

$$D_{mn\cdot}=\left(\int_0^1 (G_{\cdot 3\cdot 3}(y_3))^{-1}dy_3\right)^{-1}\int_0^1 (G_{\cdot 3\cdot 3}(y_3))^{-1}C_{mn\cdot 3}(y_3)dy_3.$$

Substituting this expression in (4.25) yields

$$\frac{dN_{\cdot mn}}{dy_3}(y_3) =$$
$$= (G_{\cdot 3 \cdot 3}(y_3))^{-1}\left[\left(\int_0^1 (G_{\cdot 3\cdot 3}(y_3))^{-1}dy_3\right)^{-1}\int_0^1 (G_{\cdot 3\cdot 3}(y_3))^{-1}G_{mn\cdot 3}(y_3)dy_3 \;-\; G_{mn\cdot 3}(y)\right]. \tag{4.26}$$

We exploit now the fact that the cell consists of a finite number of sublayers described by constant tensors and therefore the tensor $G_{mnkl}(y_3)$ is piecewise constant. Denoting by $N^s_{\cdot mn}(y_3)$ the restriction of $N_{\cdot mn}(y_3)$ to the s-th sublayer we can rewrite (4.26) for the s-th sublayer as follows:

$$\frac{dN^s_{\cdot mn}}{dy_3}(y_3) = (G^s_{\cdot 3\cdot 3})^{-1}\left[\left(\sum_r h^r (G^r_{\cdot 3\cdot 3})^{-1}\right)^{-1}\sum_r h^r (G^r_{\cdot 3\cdot 3})^{-1}G^r_{mn\cdot 3} \;-\; G^s_{mn\cdot 3}\right], \tag{4.27}$$

where G^s and G^r are the elasticity tensors for the s'th and r'th sublayers respectively; h^r is a normalized thickness of the r'th sublayer such that $\sum\limits_r h^r = 1$. All the summations are taken over the sublayers. Note that the right-hand side of (4.27) does not depend on y_3. This implies that the functions $\dfrac{dN^s_{imn}}{dy_3}(y_3)$ are constant and functions $N_{\cdot mn}$ are piecewise linear. Denoting

$$\widehat{N}^s_{\cdot mn} := \frac{dN^s_{\cdot mn}}{dy_3}(y_3),$$

we rewrite the expression (4.??) for the tensor G as follows:

$$G_{mnkl} = \sum_s h^s G^s_{mnkl} + \sum_s h^s C^s_{i3kl}\widehat{N}^s_{imn}.$$

We formulate now this final result as a theorem.

Theorem 4.13. *Suppose that the periodic cell is composed of M sublayers characterized by the stiffness tensors $G^1, ..., G^M$ with the relative thicknesses $h^1, ..., h^M$. Then the homogenized material is characterized by the stiffness tensor $G^{\mathtt{hom}}$ that is determined by*

$$G^{\mathtt{hom}}_{mnkl} = \sum_{s=1}^M h^s G^s_{mnkl} + \sum_{s=1}^M h^s G^s_{i3kl}\widehat{N}^s_{imn}, \tag{4.28}$$

where

$$\widehat{N}^{s}_{\cdot mn} = (G^{s}_{\cdot 3\cdot 3})^{-1}\left[\left(\sum_{r=1}^{M} h^{r}(G^{r}_{\cdot 3\cdot 3})^{-1}\right)^{-1}\sum_{r=1}^{M} h^{r}(G^{r}_{\cdot 3\cdot 3})^{-1}G^{r}_{mn\cdot 3} - G^{s}_{mn\cdot 3}\right]. \tag{4.29}$$

4.5 Rate of Convergence

Throughout this section we denote by u^ε the sequence of solutions of Problem 4.1 as $\varepsilon \to 0$ and by $u^0 \in H^1_0(\Omega;\mathbb{R}^3)$ and $u^1 \in \left[L^2(\Omega; H^1_\#(Y))\right]^3$ the limiting functions as described in Theorem 4.8. In that theorem we established the following convergence results:

$$\begin{aligned} u^\varepsilon &\rightharpoonup u^0 && \text{in } H^1(\Omega;\mathbb{R}^3),\\ \nabla u^\varepsilon &\underset{2}{\rightharpoonup} \nabla u^0 + \nabla_y u^1. \end{aligned}$$

Since $H^1(\Omega;\mathbb{R}^3)$ is compactly embedded in $L^2(\Omega;\mathbb{R}^3)$, the first result implies that

$$u^\varepsilon \to u^0 \qquad \text{strongly in } L^2(\Omega;\mathbb{R}^3).$$

Besides, it implies the weak convergence for the gradients, that is,

$$\nabla u^\varepsilon \rightharpoonup \nabla u^0 \qquad \text{weakly in } L^2(\Omega;\mathbb{R}^{3\times 3}).$$

In order to get the strong convergence here, we have to add an extra term, the so called *corrector*. Put

$$u^{1\varepsilon}(x) := u^0(x) + \varepsilon u^1\left(x, \frac{x}{\varepsilon}\right).$$

We will refer to this function as *the first approximation* of u^ε. Using (4.11), it can be rewritten as

$$u^{1\varepsilon}(x) = u^0(x) + \varepsilon N_{\cdot mn}\left(\frac{x}{\varepsilon}\right)\frac{\partial u^0_m}{\partial x_n}(x). \tag{4.30}$$

Recall that by definition $N_{\cdot mn} \in H^1_\#(Y;\mathbb{R}^3)$ and $\dfrac{\partial u^0_m}{\partial x_n}(x) \in L^2(\Omega)$. Hence the products $N_{\cdot mn}\left(\dfrac{x}{\varepsilon}\right)\dfrac{\partial u^0_m}{\partial x_n}(x)$ may not be weakly differentiable and $u^{1\varepsilon}$ does not necessarily lie in $H^1(\Omega;\mathbb{R}^3)$. But under some additional assumptions $u^{1\varepsilon}$ does belong to $H^1(\Omega;\mathbb{R}^3)$ and it can be shown that

$$\left\|u^\varepsilon - u^{1\varepsilon}\right\|_{H^1(\Omega;\mathbb{R}^3)} \to 0.$$

In this section we investigate the conditions sufficient for $u^{1\varepsilon}$ to be in $H^1(\Omega;\mathbb{R}^3)$ and estimate the rate of the convergences in H^1.

For the sake of convenience let us put here again the limiting equations for u^0 and $N_{\cdot mn}$. As follows from Theorem 4.8, $u^0 \in H_0^1(\Omega;\mathbb{R}^3)$ is a unique solution of the homogenized equation

$$\int_\Omega G^{\mathbf{hom}}_{mnkl} \frac{\partial u^0_m}{\partial x_n}(x) \frac{\partial v^0_k}{\partial x_l}(x)\, dx = \int_\Omega f_k v^0_k \, dx \qquad \forall v^0 \in H_0^1(\Omega;\mathbb{R}^3), \tag{4.31}$$

$N_{\cdot mn} \in H^1_\#(Y;\mathbb{R}^3)$ are unique solutions of the cell equations

$$\int_Y G_{ijkl}(y) \frac{\partial N_{imn}}{\partial y_j}(y) \frac{\partial v^1_k}{\partial y_l}(y)\, dy = -\int_Y G_{mnkl}(y) \frac{\partial v^1_k}{\partial y_l}(y)\, dy \quad \forall v^1 \in H^1_\#(Y;\mathbb{R}^3). \tag{4.32}$$

So far we have not made any assumptions about the smoothness of $\partial\Omega$; it only had to be Lipschitz. From now on we assume that $\partial\Omega$ is of class $\mathcal{C}^2$. This enables us to increase the regularity of the solution to (4.31). The following theorem holds:

Theorem 4.14. *Let Ω be a domain in $\mathbb{R}^3$ with a boundary $\partial\Omega$ of class $\mathcal{C}^2$, let $f \in L^2(\Omega;\mathbb{R}^3)$. Then the solution u^0 to (4.31) belongs to the space $H^2(\Omega;\mathbb{R}^3)$.*

This theorem can be found for example in [10] (see Theorem 6.3-6).

Hence $u^0 \in H^2(\Omega;\mathbb{R}^3)$. However, this is still not enough for $u^{1\varepsilon}$ to be in $H^1(\Omega;\mathbb{R}^3)$. We need also a higher regularity of $N_{\cdot mn}$. It is shown in [56] that a higher regularity takes place if the boundaries between the materials composing the unit cell are smooth. More formally, denote by $Y_1, ...Y_M$ the open subdomans of Y corresponding to the composing materials $1, ...M$ and repeated periodically in all the direction such that

$$\bigcup_{i=1}^M \overline{Y}_i = \mathbb{R}^3, \qquad Y = \bigcup_{i=1}^M \overline{Y_i \cap Y}.$$

Following [56], we say that a Y-periodic function *belongs to class $\hat{\mathcal{C}}$* if it has bounded derivatives of any order in $Y_i, i = 1, ..., M$. Note that the tensor function $G(y)$ belongs to $\hat{\mathcal{C}}$ because it is constant in any Y_i. The following result is due to [56] (Theorem 6.2, Chapter I).

Theorem 4.15. *Let $w \in H^1_\#(Y;\mathbb{R}^3)$ be a solution of*

$$\int_Y A_{ijkl}(y) \frac{\partial w_i}{\partial y_j}(y) \frac{\partial v^1_k}{\partial y_l}(y)\, dy = -\int_Y F_{kl}(y) \frac{\partial v^1_k}{\partial y_l}(y)\, dy \qquad \forall v^1 \in H^1_\#(Y;\mathbb{R}^3),$$

where $A_{ijkl}(y)$ and $F_{kl}(y)$ belong to $\hat{\mathcal{C}}$.
Further, suppose that the boundaries $\partial Y_1, ..., \partial Y_M$ are smooth.

Then w also belongs to $\hat{\mathcal{C}}$, that is, w is piecewise smooth with bounded derivatives in any Y_i, $i = 1, ..., M$.

Combining Theorems 4.14 and 4.15, we discover that if $\partial\Omega$ and ∂Y_i are smooth, then $u^0 \in H^2(\Omega; \mathbb{R}^3)$ and $N_{\cdot mn} \in \hat{\mathcal{C}}$. Hence $u^{1\varepsilon} \in H^1(\Omega; \mathbb{R}^3)$. Under these assumptions the following estimate can be shown.

Theorem 4.16. *Let Ω, $\{Y_i\}_{i=1}^M$, u^ε, u^0, $u^{1\varepsilon}$ and f be defined as above. Suppose that*

1. *$\partial\Omega$ is smooth,*

2. *∂Y_i are smooth for all $i = 1, ..., M$,*

3. *$f \in H^1(\Omega; \mathbb{R}^3)$.*

Then there is a positive constant C such that

$$\left\| u^\varepsilon - u^{1\varepsilon} \right\|_{H^1(\Omega;\mathbb{R}^3)} < C\sqrt{\varepsilon}.$$

The theorem is proved in [56] (Theorem 1.2, Chapter II). However, the condition $f \in H^1(\Omega; \mathbb{R}^3)$ is seldom fulfilled in real applications. We derive now an estimate for the case where $f \in L^2(\Omega; \mathbb{R}^3)$. Besides, we also drop the requirement of smoothness of the boundaries ∂Y_i replacing it by a weaker condition. Before formulating the main result, we first prove some auxiliary propositions.

Lemma 4.17 (Stampacchia). *Let φ be a nonnegative, nonincreasing function defined on $[c_0, \infty)$. Suppose that there exist constants $C > 0, \alpha > 0$, and $\beta > 1$ such that*

$$\varphi(d) \leqslant \frac{C}{(d-c)^\alpha} \varphi(c)^\beta$$

for all $d > c \geqslant c_0$.
Then there exist c_1 such that

$$\varphi(c) = 0 \qquad \forall c \geqslant c_1.$$

The lemma is due to Stampacchia (see [67], Lemma 4.1).

Theorem 4.18. *Let $A \in \left(L^\infty_\#(Y)\right)^{3\times 3}$ be a positive definite and symmetric matrix for almost all $y \in Y$, $g \in L^2_\#(Y;\mathbb{R}^3)$. Then the problem*

$$\int\limits_Y A(y)\nabla u(y)\cdot\nabla\varphi(y)\,dy = \int\limits_Y g(y)\cdot\nabla\varphi(y)\,dy \qquad \forall\varphi \in H^1_\#(Y) \tag{4.33}$$

is uniquely solvable in $H^1_\#(Y)$.
Futhermore, if $g \in L^\infty_\#(Y;\mathbb{R}^3)$, then $u \in L^\infty_\#(Y)$.

Proof. The solvability and uniqueness of (4.33) in $H^1_\#(Y)$ is due to the Lax-Milgram theorem and the Poincaré inequality. To prove that $g \in L^\infty_\#(Y;\mathbb{R}^3)$ implies $u \in L^\infty_\#(Y)$, we exploit the ideas from [12].

Suppose that $g \in L^\infty_\#(Y;\mathbb{R}^3)$ and $u \in H^1_\#(Y)$ is the unique solution of (4.33). Let c be an arbitrary positive number from $\mathbb{R}$. Let us define

$$u_c(y) := \begin{cases} u(y)-c & \text{if } u(y) \geqslant c, \\ 0 & \text{if } |u(y)| < c, \\ u(y)+c & \text{if } u(y) \leqslant -c. \end{cases}$$

We would like to show now that $u_c \in H^1(Y)$. This is not obviously and can not be shown straightforwardly because the sets

$$\{y \in Y \,:\, u(y) \geqslant c\}, \quad \{y \in Y \,:\, |u(y)| < c\}, \quad \{y \in Y \,:\, u(y) \leqslant -c\}$$

do not have to possess a piecewise smooth boundary and we can not integrate by parts. Instead of it, we exploit a roundabout characterization of H^1-functions. It is known (see [55] and [51], § 1.1.3) that the restriction of a H^1-function to almost every line parallel to the coordinate directions, possibly after modifying the function on a set of measure zero, is absolutely continuous. This means that the pointwise gradient exists almost everywhere. Moreover, it agrees with the weak gradient. Then the pointwise gradient of u_c also exists almost everywhere and

$$\nabla u_c(y) = \begin{cases} \nabla u(y) & \text{if } |u(y)| \geqslant c \\ 0 & \text{if } |u(y)| < c. \end{cases}$$

Obviously, ∇u_c belongs to $L^2(Y;\mathbb{R}^3)$ and u_c belongs to $L^2(Y)$. This implies that $u_c \in H^1(Y)$. Moreover, u_c is Y-periodic and hence $u_c - \langle u_c\rangle_Y \in H^1_\#(Y)$.

Let us further define

$$S_c(u) := \{y \in Y : |u(y)| \geqslant c\}.$$

Note that $S_c(u)$ is measurable. For the sake of brevity, we will omit the argument, writing just S_c. Since $u_c - \langle u_c\rangle_Y \in H^1_\#(Y)$, it can be used as a test function in (4.33). This yields

$$\int_Y A\nabla u \cdot \nabla u_c \, dy = \int_Y g \cdot \nabla u_c \, dy = \int_{S_c} g \cdot \nabla u_c \leqslant$$

$$\leqslant \|g\|_{L^2(S_c;\mathbb{R}^3)} \|\nabla u_c\|_{L^2(S_c;\mathbb{R}^3)} \leqslant C|S_c|^{\frac{1}{2}} \|\nabla u_c\|_{L^2(Y;\mathbb{R}^3)}$$

On the other hand, since A is positive definite, we have

$$\int_Y A\nabla u \cdot \nabla u_c \, dy = \int_{S_c} A\nabla u_c \cdot \nabla u_c \, dy = \int_Y A\nabla u_c \cdot \nabla u_c \, dy \geqslant C\|\nabla u_c\|^2_{L^2(Y;\mathbb{R}^3)}.$$

Combining the last two inequalities, we get:

$$|S_c| \geqslant C\|\nabla u_c\|^2_{L^2(Y;\mathbb{R}^3)}. \tag{4.34}$$

We exploit the fact that $H^1(Y)$ is continuously embedded in $L^4(Y)$ (see [2]) and derive an estimate for $\|\nabla u_c\|^2_{L^2(Y;\mathbb{R}^3)}$ as follows:

$$\|u_c - \langle u_c\rangle_Y\|^2_{L^2(S_c)} = \int_{S_c} \left(u_c - \langle u_c\rangle_Y\right)^2 \, dy \leqslant$$

$$\leqslant \left(\int_{S_c} 1^2 \, dy\right)^{\frac{1}{2}} \left(\int_{S_c} \left(u_c - \langle u_c\rangle_Y\right)^4 \, dy\right)^{\frac{1}{2}} \leqslant |S_c|^{\frac{1}{2}} \|u_c - \langle u_c\rangle_Y\|^2_{L^4(Y)} \leqslant$$

$$\leqslant |S_c|^{\frac{1}{2}} C \, \|\nabla(u_c - \langle u_c\rangle_Y)\|^2_{L^2(Y;\mathbb{R}^3)} = C|S_c|^{\frac{1}{2}} \|\nabla u_c\|^2_{L^2(Y;\mathbb{R}^3)}.$$

Substituting this result into (4.34), we obtain

$$C|S_c|^{\frac{3}{2}} \geqslant \|u_c - \langle u_c\rangle_Y\|^2_{L^2(S_c)} = \int_{S_c} \left(u_c - \langle u_c\rangle_Y\right)^2 \, dy =$$

$$= \int_{S_c} u_c^2 \, dy - 2\langle u_c\rangle_Y \int_{S_c} u_c \, dy + \int_{S_c} \langle u_c\rangle_Y^2 \, dy =$$

$$= \|u_c\|^2_{L^2(S_c)} - 2\langle u_c\rangle_Y \int\limits_Y u_c\,dy + \int\limits_{S_c} \langle u_c\rangle_Y^2\,dy \geqslant$$

$$\geqslant \|u_c\|^2_{L^2(S_c)} - 2\langle u_c\rangle_Y^2,$$

which implies

$$C|S_c|^{\frac{3}{2}} + 2\langle u_c\rangle_Y^2 \geqslant \|u_c\|^2_{L^2(S_c)}. \tag{4.35}$$

The next property follows from the definition of u_c and S_c:

$$\forall d > c \quad \|u_c\|^2_{L^2(S_c)} \geqslant \|u_c\|^2_{L^2(S_d)} \geqslant (d-c)^2|S_d|.$$

Combining this with (4.35) yields

$$\forall d > c \quad |S_h| \leqslant \frac{C|S_c|^{\frac{3}{2}} + \langle u_c\rangle_Y^2}{(d-c)^2}. \tag{4.36}$$

We derive now an estimate for $\langle u_c\rangle_Y$ by the Hölder inequality (recall that $u_c \in L^4(Y)$ due to the Sobolev embedding). We have

$$\langle u_c\rangle_Y = \int\limits_Y u_c\,dy = \int\limits_{S_c} u_c\,dy \leqslant \left(\int\limits_{S_c} 1^{\frac{4}{3}}\,dy\right)^{\frac{3}{4}} \left(\int\limits_{S_c} u_c^4\,dy\right)^{\frac{1}{4}} \leqslant$$

$$\leqslant |S_c|^{\frac{3}{4}} \left(\int\limits_Y u_c^4\,du\right)^{\frac{1}{4}} = |S_c|^{\frac{3}{4}}\|u_c\|_{L^4(Y)} \leqslant C|S_c|^{\frac{3}{4}}\|u_c\|_{H^1(Y)} \leqslant C|S_c|^{\frac{3}{4}}\|u\|_{H^1_\#(Y)}.$$

Substituting this estimate in (4.36), we finally get

$$\forall d > c \quad |S_d| \leqslant \frac{C\|u\|_{H^1_\#(Y)}}{(d-c)^2}|S_c|^{\frac{3}{2}}. \tag{4.37}$$

Let us now consider the mapping $c \mapsto |S_c|$. This is a nonnegative nonincreasing real function. Using Lemma 4.17, we obtain that there exists c_0 such that

$$\forall c > c_0 \quad |S_c(u)| = 0.$$

This means that $\|u\|_{L^\infty_\#(Y)} < \infty$.

□

Theorem 4.19. *Let $g \in L^2_\#(Y;\mathbb{R}^3)$ be a vector field satisfying*

$$\int_Y g_l(y)\frac{\partial \varphi}{\partial y_l}(y)\,dy = 0 \qquad \forall \varphi \in H^1_\#(Y). \tag{4.38}$$

Then there exists a skew-symmetric matrix $\alpha \in H^1_\#(Y;\mathbb{R}^{3\times 3})$ such that

$$g_j = \langle g_j\rangle_Y + \frac{\partial \alpha_{ij}}{\partial y_i}.$$

Moreover, if $g \in L^\infty_\#(Y;\mathbb{R}^3)$ then $\alpha \in L^\infty_\#(Y;\mathbb{R}^{3\times 3})$.

Proof. The first part is based on the following result from [33]. It is proven there (p. 6–7) that if $g \in L^2_\#(Y;\mathbb{R}^3)$ satisfies (4.38), then it admits the representation

$$g_j = \langle g_j\rangle_Y + \frac{\partial \alpha_{ij}}{\partial y_i},$$

where $\alpha_{ij} \in H^1_\#(Y)$, $\langle \alpha_{ij}\rangle_Y = 0$, $\alpha_{ij} = -\alpha_{ji}$. We define now a vector potential h by

$$h := (\alpha_{23},\, \alpha_{31},\, \alpha_{12})^T.$$

Obviously $h \in H^1_\#(Y;\mathbb{R}^3)$. It can be verified directly that

$$\frac{\partial \alpha_{ij}}{\partial y_i} = (\operatorname{curl} h)_j$$

and hence

$$g = \langle g\rangle_Y + \operatorname{curl} h. \tag{4.39}$$

Moreover, by construction of α_{ij} (see [33], p. 6–7 for details), it can be checked straightforwardly that

$$\operatorname{div} h = 0.$$

Let us now show that $g \in L^\infty_\#(Y;\mathbb{R}^3)$ implies $h \in L^\infty_\#(Y;\mathbb{R}^3)$. Let φ be an arbitrary vector function from $\varphi \in C^\infty_\#(Y;\mathbb{R}^3)$. Multiplying (4.39) by $\operatorname{curl}\varphi$ and integrating over Y, we obtain

$$\int_Y (g - \langle g\rangle_Y)\cdot \operatorname{curl}\varphi\,dy = \int_Y \operatorname{curl} h \cdot \operatorname{curl}\varphi\,dy =$$

$$= \int_Y h \cdot \operatorname{curl}(\operatorname{curl}\varphi)\,dy + \int_{\partial Y} (\nu \times h)\cdot \operatorname{curl}\varphi\,ds = \int_Y h\cdot(-\Delta\,\varphi + \nabla \operatorname{div}\varphi)\,dy =$$

$$= \int_Y \nabla h : \nabla\varphi\,dy - \int_{\partial Y} h\cdot\nabla\varphi\nu\,ds - \int_Y \operatorname{div} h \operatorname{div}\varphi\,dy + \int_{\partial Y} \operatorname{div}\varphi\, h\cdot\nu\,ds =$$

$$= \int_Y \nabla h : \nabla\varphi\,dy,$$

where ν is the outward unit normal to ∂Y. The boundary terms in the derivation above vanish because h and φ are Y-periodic. We have then that h satisfies

$$\int_Y \nabla h : \nabla\varphi\,dy, = \int_Y (g - \langle g\rangle_Y)\cdot \operatorname{curl}\varphi\,dy \tag{4.40}$$

for all $\varphi \in \mathcal{C}^\infty_\#(Y;\mathbb{R}^3)$. By definition $\mathcal{C}^\infty_\#(Y;\mathbb{R}^3)$ is dense in $H^1_\#(Y;\mathbb{R}^3)$ and hence (4.40) holds also for all $\varphi \in H^1_\#(Y;\mathbb{R}^3)$. Taking $\varphi = (\varphi_1, 0, 0)^T$ we obtain

$$\int_Y \nabla h_1 \cdot \nabla\varphi_1\,dy, = \int_Y (g - \langle g\rangle_Y)\cdot\left(0, -\frac{\partial\varphi_1}{\partial y_3}, \frac{\partial\varphi_1}{\partial y_2}\right)^T dy \qquad \forall\varphi_1 \in H^1_\#(Y).$$

This is equivalent to

$$\int_Y \nabla h_1 \cdot \nabla\varphi_1\,dy, = \int_Y (0,\ g_3 - \langle g_3\rangle_Y,\ -g_2 + \langle g_2\rangle_Y)^T \cdot \nabla\varphi_1\,dy \qquad \forall\varphi_1 \in H^1_\#(Y).$$

By Theorem 4.18 this implies that $h_1 \in L^\infty_\#(Y)$. Similarly, taking $\varphi = (0, \varphi_2, 0)$ and $\varphi = (0, 0, \varphi_3)$, we discover that $h_2 \in L^\infty_\#(Y)$ and $h_3 \in L^\infty_\#(Y)$. Hence, by definition of h, we have that $\alpha_{12}, \alpha_{13}, \alpha_{23} \in L^\infty_\#(Y)$. Since the matrix α is skew-symmetric, this means that $\alpha \in L^\infty_\#(Y;\mathbb{R}^{3\times3})$.

□

Theorem 4.20 (Estimate in $H^1(\Omega;\mathbb{R}^3)$).

Let Ω be an open bounded Lipschitz domain, and let u^ε, u^0, $N_{\cdot mn}$, $u^{1\varepsilon}$, and f be defined as above. Suppose that

1. *$\partial\Omega$ is of class $\mathcal{C}^2$,*
2. *$\forall\, m, n \quad N_{\cdot mn} \in W^{1,\infty}_\#(Y;\mathbb{R}^3)$,*
3. *$f \in L^2(\Omega;\mathbb{R}^3)$.*

Then there is a positive constant C such that

$$\left\| u^{\varepsilon} - u^{1\varepsilon} \right\|_{H^1(\Omega;\mathbb{R}^3)} < C\sqrt[3]{\varepsilon}. \tag{4.41}$$

Proof. The proof partly repeats the proof of Theorem 4.16 with necessary modifications caused by the weaker regularity assumptions. We exploited also some ideas from [33] (Section I.4).

First note that since $N_{\cdot mn} \in W^{1,\infty}_{\#}(Y;\mathbb{R}^3)$ and $u^0 \in H^2(\Omega;\mathbb{R}^3)$, the first approximation

$$u^{1\varepsilon}(x) = u^0(x) + \varepsilon N_{\cdot mn}\left(\frac{x}{\varepsilon}\right)\frac{\partial u^0_m}{\partial x_n}(x)$$

belongs to $H^1(\Omega;\mathbb{R}^3)$ and for all i, j

$$\frac{\partial u^{1\varepsilon}_i}{\partial x_j}(x) = \frac{\partial u^0_i}{\partial x_j}(x) + \frac{\partial N_{imn}}{\partial y_j}\left(\frac{x}{\varepsilon}\right)\frac{\partial u^0_m}{\partial x_n}(x) + \varepsilon N_{imn}\left(\frac{x}{\varepsilon}\right)\frac{\partial^2 u^0_m}{\partial x_n \partial x_j}(x),$$

where $y := \varepsilon^{-1}x$. Let the form $a^{\varepsilon}(\cdot,\cdot)$ be given by (4.3). We take now an arbitrary $v \in H^1_0(\Omega;\mathbb{R}^3)$ and derive an estimate for $a^{\varepsilon}(u^{\varepsilon} - u^{1\varepsilon}, v)$. We have

$$a^{\varepsilon}(u^{\varepsilon} - u^{1\varepsilon}, v) = \int_{\Omega} G_{ijkl}\left(\frac{x}{\varepsilon}\right)\left[\frac{\partial u^{\varepsilon}_i}{\partial x_j}(x) - \frac{\partial u^{1\varepsilon}_i}{\partial x_j}(x)\right]\frac{\partial v_k}{\partial x_l}(x)\,dx =$$

$$= \int_{\Omega} G_{ijkl}\left(\frac{x}{\varepsilon}\right)\frac{\partial u^{\varepsilon}_i}{\partial x_j}(x)\frac{\partial v_k}{\partial x_l}(x)\,dx - \int_{\Omega} G_{ijkl}\left(\frac{x}{\varepsilon}\right)\frac{\partial u^{1\varepsilon}_i}{\partial x_j}(x)\frac{\partial v_k}{\partial x_l}(x)\,dx =$$

$$= \int_{\Omega} G^{\mathbf{hom}}_{ijkl}\frac{\partial u^0_i}{\partial x_j}(x)\frac{\partial v_k}{\partial x_l}(x)\,dx - \int_{\Omega} G_{ijkl}\left(\frac{x}{\varepsilon}\right)\frac{\partial u^{1\varepsilon}_i}{\partial x_j}(x)\frac{\partial v_k}{\partial x_l}(x)\,dx =$$

$$= \int_{\Omega}\left[G^{\mathbf{hom}}_{ijkl}\frac{\partial u^0_i}{\partial x_j}(x) - G_{ijkl}\left(\frac{x}{\varepsilon}\right)\left(\frac{\partial u^0_i}{\partial x_j}(x) + \frac{\partial N_{imn}}{\partial y_j}\left(\frac{x}{\varepsilon}\right)\frac{\partial u^0_m}{\partial x_n}(x)\right)\right]\frac{\partial v_k}{\partial x_l}(x)\,dx +$$

$$+\varepsilon\int_{\Omega} G_{ijkl}\left(\frac{x}{\varepsilon}\right)N_{imn}\left(\frac{x}{\varepsilon}\right)\frac{\partial^2 u^0_m}{\partial x_n \partial x_j}(x)\frac{\partial v_k}{\partial x_l}(x)\,dx =$$

$$= \int_\Omega \frac{\partial u_m^0}{\partial x_n}(x) \left[G_{mnkl}^{\mathrm{hom}} - G_{mnkl}\left(\frac{x}{\varepsilon}\right) - G_{ijkl}\left(\frac{x}{\varepsilon}\right) \frac{\partial N_{imn}}{\partial y_j}\left(\frac{x}{\varepsilon}\right)\right] \frac{\partial v_k}{\partial x_l}(x)\, dx +$$

$$+\varepsilon \int_\Omega G_{ijkl}\left(\frac{x}{\varepsilon}\right) N_{imn}\left(\frac{x}{\varepsilon}\right) \frac{\partial^2 u_m^0}{\partial x_n \partial x_j}(x) \frac{\partial v_k}{\partial x_l}(x)\, dx =$$

$$= \int_\Omega \frac{\partial u_m^0}{\partial x_n}(x) g_l^{mnk}\left(\frac{x}{\varepsilon}\right) \frac{\partial v_k}{\partial x_l}(x)\, dx + \varepsilon \int_\Omega G_{ijkl}\left(\frac{x}{\varepsilon}\right) N_{imn}\left(\frac{x}{\varepsilon}\right) \frac{\partial^2 u_m^0}{\partial x_n \partial x_j}(x) \frac{\partial v_k}{\partial x_l}(x)\, dx,$$

where

$$g_l^{mnk}(y) := G_{mnkl}^{\mathrm{hom}} - G_{mnkl}(y) - G_{ijkl}(y) \frac{\partial N_{imn}}{\partial y_j}(y).$$

Note that $\forall m, n, k, l \;\; g_l^{mnk} \in L_\#^\infty(Y)$ because $N_{\cdot mn} \in W_\#^{1,\infty}(Y; \mathbb{R}^3)$ and $G_{mnkl} \in L_\#^\infty(Y)$. Moreover, due to (4.32),

$$\int_Y g_l^{mnk} \frac{\partial v}{\partial y_l}\, dy = 0 \qquad \forall v \in H_\#^1(Y) \quad \forall m, n, k,$$

and, by definition of G^{hom} (see (4.14)),

$$\langle g_l^{mnk} \rangle_Y = 0 \qquad \forall m, n, k, l.$$

Then by Theorem 4.19, g_l^{mnk} admits the representation

$$g_l^{mnk} = \frac{\partial \alpha_{lr}^{mnk}}{\partial y_r} \qquad \forall m, n, k, l, \tag{4.42}$$

where the matrix $\alpha_{\cdot\cdot}^{mnk} \in H_\#^1(Y; \mathbb{R}^{3\times3}) \cap L_\#^\infty(Y; \mathbb{R}^{3\times3})$ is skew-symmetric for all m, n, k, that is,

$$\alpha_{lr}^{mnk} = -\alpha_{rl}^{mnk}.$$

Moreover, (4.42) implies that the sum $\dfrac{\partial \alpha_{lr}^{mnk}}{\partial y_r} \in L_\#^\infty(Y)$. We can then derive that for all m, n, k, l the product $\dfrac{\partial u_m^0}{\partial x_n}(x) \alpha_{l\cdot}^{mnk}\left(\dfrac{x}{\varepsilon}\right)$ belongs to $H(\mathrm{div}; \Omega)$ and

$$\frac{\partial}{\partial x_r}\left(\frac{\partial u_m^0}{\partial x_n}(x) \alpha_{lr}^{mnk}\left(\frac{x}{\varepsilon}\right)\right) = \frac{\partial^2 u_m^0}{\partial x_n \partial x_r}(x) \alpha_{lr}^{mnk}\left(\frac{x}{\varepsilon}\right) + \frac{1}{\varepsilon} \frac{\partial u_m^0}{\partial x_n}(x) \frac{\partial \alpha_{lr}^{mnk}}{\partial y_r}\left(\frac{x}{\varepsilon}\right) \in L^2(\Omega).$$

Hence,

$$\frac{\partial u_m^0}{\partial x_n}(x) g_l^{mnk}\left(\frac{x}{\varepsilon}\right) = \varepsilon \frac{\partial}{\partial x_r}\left(\frac{\partial u_m^0}{\partial x_n}(x)\alpha_{lr}^{mnk}\left(\frac{x}{\varepsilon}\right)\right) - \varepsilon \frac{\partial^2 u_m^0}{\partial x_n \partial x_r}(x)\alpha_{lr}^{mnk}\left(\frac{x}{\varepsilon}\right).$$

Substituting this relation into the expression for $a^\varepsilon(u^\varepsilon - u^{1\varepsilon}, v)$ above yields

$$\begin{aligned} a^\varepsilon(u^\varepsilon - u^{1\varepsilon}, v) &= \\ = \varepsilon \int\limits_\Omega \frac{\partial}{\partial x_r}\left(\frac{\partial u_m^0}{\partial x_n}(x)\alpha_{lr}^{mnk}\left(\frac{x}{\varepsilon}\right)\right)\frac{\partial v_k}{\partial x_l}(x)\,dx &- \varepsilon \int\limits_\Omega \frac{\partial^2 u_m^0}{\partial x_n \partial x_r}(x)\alpha_{lr}^{mnk}\left(\frac{x}{\varepsilon}\right)\frac{\partial v_k}{\partial x_l}(x)\,dx + \\ &+\varepsilon \int\limits_\Omega G_{ijkl}\left(\frac{x}{\varepsilon}\right) N_{imn}\left(\frac{x}{\varepsilon}\right)\frac{\partial^2 u_m^0}{\partial x_n \partial x_j}(x)\frac{\partial v_k}{\partial x_l}(x)\,dx. \end{aligned} \tag{4.43}$$

Let us consider the first term on the right-hand side. Recall that $\frac{\partial u_m^0}{\partial x_n}(x)\alpha_{l\cdot}^{mnk}\left(\frac{x}{\varepsilon}\right) \in H(\mathrm{div};\Omega)$. Since $\alpha_{lr}^{mnk} = -\alpha_{rl}^{mnk}$, we have that $\frac{\partial u_m^0}{\partial x_n}(x)\alpha_{\cdot r}^{mnk}\left(\frac{x}{\varepsilon}\right)$ also belongs to $H(\mathrm{div};\Omega)$. Then by the Green theorem for functions from $H(\mathrm{div};\Omega)$, we obtain

$$\begin{aligned} &\forall \varphi \in C_0^\infty(\Omega) \quad \forall m,n,k \\ &\int\limits_\Omega \frac{\partial}{\partial x_r}\left(\frac{\partial u_m^0}{\partial x_n}(x)\alpha_{lr}^{mnk}\left(\frac{x}{\varepsilon}\right)\right)\frac{\partial \varphi}{\partial x_l}(x)\,dx = \int\limits_\Omega \left(\frac{\partial u_m^0}{\partial x_n}(x)\alpha_{lr}^{mnk}\left(\frac{x}{\varepsilon}\right)\right)\frac{\partial^2 \varphi}{\partial x_l \partial x_r}(x)\,dx = \\ &\qquad = \int\limits_\Omega \frac{\partial}{\partial x_l}\left(\frac{\partial u_m^0}{\partial x_n}(x)\alpha_{lr}^{mnk}\left(\frac{x}{\varepsilon}\right)\right)\frac{\partial \varphi}{\partial x_r}(x)\,dx. \end{aligned}$$

Since $C_0^\infty(\Omega)$ is dense in $H_0^1(\Omega)$, the equailty above holds also for $\varphi \in H_0^1(\Omega)$. Therefore, for the first term on the right-hand side of (4.43) we have

$$\int\limits_\Omega \frac{\partial}{\partial x_r}\left(\frac{\partial u_m^0}{\partial x_n}(x)\alpha_{lr}^{mnk}\left(\frac{x}{\varepsilon}\right)\right)\frac{\partial v^k}{\partial x_l}(x)\,dx = \int\limits_\Omega \frac{\partial}{\partial x_l}\left(\frac{\partial u_m^0}{\partial x_n}(x)\alpha_{lr}^{mnk}\left(\frac{x}{\varepsilon}\right)\right)\frac{\partial v^k}{\partial x_r}(x)\,dx.$$

Swapping the indices l and r and putting $\alpha_{rl}^{mnk} = -\alpha_{lr}^{mnk}$ on the right-hand side, we discover that the first term on the fight-hand side of (4.43) must be zero. The other two terms are estimated by the Cauchy-Schwarz inequality. Finally we obtain

$$\left|a^\varepsilon(u^\varepsilon - u^{1\varepsilon}, v)\right| \leqslant C\varepsilon \|v\|_{H^1(\Omega;\mathbb{R}^3)}. \tag{4.44}$$

If the difference $u^\varepsilon - u^{1\varepsilon}$ were in $H_0^1(\Omega;\mathbb{R}^3)^3$, we could substitute it into the estimate above in place of v and derive an estimate for $\|u^\varepsilon - u^{1\varepsilon}\|_{H^1(\Omega;\mathbb{R}^3)}$ by the ellipticity of a^ε. Unfortunately $u^{1\varepsilon}$ does not have to vanish on $\partial\Omega$. In order to avoid this difficulty, we construct an auxiliary function $w^\varepsilon \in H_0^1(\Omega;\mathbb{R}^3)$ that approximates $u^{1\varepsilon}$, and derive estimates for the differences $u^\varepsilon - w^\varepsilon$ and $w^\varepsilon - u^{1\varepsilon}$.

Denote by Ω_ε the internal ε-neighborhood of $\partial\Omega$, that is,

$$\Omega_\varepsilon := \{x \in \Omega \mid \rho(x, \partial\Omega) < \varepsilon\},$$

where $\rho(x, \partial\Omega)$ is the distance from x to $\partial\Omega$. Further, let τ^ε be a family of cutoff functions from $C_0^\infty(\Omega)$ satisfying

1. $0 \leqslant \tau^\varepsilon \leqslant 1$, $\tau^\varepsilon \equiv 1$ on $\Omega \setminus \Omega_\varepsilon$,
2. $\varepsilon\,|\nabla\tau^\varepsilon| \leqslant C$, where C does not depend on ε.

A family of functions with these properties can always be constructed (see [29], Theorem 1.4.2). We define now

$$w_i^\varepsilon(x) := u_i^0(x) + \varepsilon\tau^\varepsilon(x) N_{imn}\left(\frac{x}{\varepsilon}\right)\frac{\partial u_m^0}{\partial x_n}(x)$$

and estimate the difference $w^\varepsilon - u^{1\varepsilon}$ as follows:

$$w^\varepsilon - u^{1\varepsilon} = \varepsilon(\tau^\varepsilon(x) - 1)N_{\cdot mn}\left(\frac{x}{\varepsilon}\right)\frac{\partial u_m^0}{\partial x_n}(x),$$

$$\begin{aligned}\frac{\partial}{\partial x_j}\left(w^\varepsilon - u^{1\varepsilon}\right) = &\varepsilon\frac{\partial\tau^\varepsilon}{\partial x_j}(x)N_{\cdot mn}\left(\frac{x}{\varepsilon}\right)\frac{\partial u_m^0}{\partial x_n}(x)+\\ &+(\tau^\varepsilon(x)-1)\frac{\partial N_{\cdot mn}}{\partial y_j}\left(\frac{x}{\varepsilon}\right)\frac{\partial u_m^0}{\partial x_n}(x)\\ &+\varepsilon(\tau^\varepsilon(x)-1)N_{\cdot mn}\left(\frac{x}{\varepsilon}\right)\frac{\partial^2 u_m^0}{\partial x_n\partial x_j}(x).\end{aligned}$$

Since $N_{\cdot mn} \in W_\#^{1,\infty}(Y;\mathbb{R}^3)$ and the product $\varepsilon\dfrac{\partial\tau^\varepsilon}{\partial x_j}(x)$ is bounded by construction of τ^ε, we obtain

$$\|w^\varepsilon - u^{1\varepsilon}\|^2_{H^1(\Omega;\mathbb{R}^3)} \leqslant C\left(\varepsilon^2\left\|\nabla u^0\right\|^2_{L^2(\Omega_\varepsilon;\mathbb{R}^{3\times3})} + \left\|\nabla u^0\right\|^2_{L^2(\Omega_\varepsilon;\mathbb{R}^{3\times3})} + \varepsilon^2\left\|u^0\right\|^2_{H^2(\Omega;\mathbb{R}^3)}\right). \quad (4.45)$$

The first and third terms are $\mathcal{O}(\varepsilon)$ as $\varepsilon \to 0$. To get the rate of the convergence for the second term, we exploit the fact that H^1 is continuously embedded in L^6 for three-dimensional domains with a C^1 boundary. Hence, by the Hölder inequality, we have

$$\begin{aligned}\int_{\Omega_\varepsilon} \left(\frac{\partial u_m^0}{\partial x_n}\right)^2 dx \leqslant \left(\int_{\Omega_\varepsilon} 1^{\frac{3}{2}}\, dx\right)^{\frac{2}{3}} \left(\int_{\Omega_\varepsilon} \left(\frac{\partial u_m^0}{\partial x_n}\right)^6 dx\right)^{\frac{1}{3}} \leqslant |\Omega_\varepsilon|^{2/3} \left\|\frac{\partial u_m^0}{\partial x_n}\right\|^2_{L^6(\Omega)} \leqslant \\ \leqslant C\varepsilon^{2/3} \left\|\frac{\partial u_m^0}{\partial x_n}\right\|^2_{H^1(\Omega)}.\end{aligned} \tag{4.46}$$

Combining this estimate with (4.45) yields (assumed $\varepsilon < 1$)

$$\|w^\varepsilon - u^{1\varepsilon}\|_{H^1(\Omega;\mathbb{R}^3)} \leqslant C\sqrt[3]{\varepsilon}. \tag{4.47}$$

Let us now estimate the difference $u^\varepsilon - w^\varepsilon$. Using (4.44), (4.47) and the boundedness of a^ε, we get

$$\begin{aligned}a^\varepsilon(u^\varepsilon - w^\varepsilon, u^\varepsilon - w^\varepsilon) &= a^\varepsilon(u^\varepsilon - u^{1\varepsilon}, u^\varepsilon - w^\varepsilon) + a^\varepsilon(u^{1\varepsilon} - w^\varepsilon, u^\varepsilon - w^\varepsilon) \leqslant \\ &\leqslant C\varepsilon\|u^\varepsilon - w^\varepsilon\|_{H^1(\Omega;\mathbb{R}^3)} + C\|w^\varepsilon - u^{1\varepsilon}\|_{H^1(\Omega;\mathbb{R}^3)}\|u^\varepsilon - w^\varepsilon\|_{H^1(\Omega;\mathbb{R}^3)} \leqslant \\ &\leqslant C\sqrt[3]{\varepsilon}\|u^\varepsilon - w^\varepsilon\|_{H^1(\Omega;\mathbb{R}^3)}.\end{aligned}$$

Since a^ε is elliptic, this implies

$$\|u^\varepsilon - w^\varepsilon\|_{H^1(\Omega;\mathbb{R}^3)} \leqslant C\sqrt[3]{\varepsilon}.$$

Finally we obtain

$$\|u^\varepsilon - u^{1\varepsilon}\|_{H^1(\Omega;\mathbb{R}^3)} \leqslant \|u^\varepsilon - w^\varepsilon\|_{H^1(\Omega;\mathbb{R}^3)} + \|w^\varepsilon - u^{1\varepsilon}\|_{H^1(\Omega;\mathbb{R}^3)} \leqslant C\sqrt[3]{\varepsilon}.$$

This completes the proof of Theorem 4.20.

□

Remark. The power 1/3 in the final estimate arises due to the Sobolev embedding of H^1 in L^6. If u^0 possesses a higher regularity, this estimate can be improved. In particular, if $u^0 \in W^{2,p}(\Omega;\mathbb{R}^3)$ for some $p > 3$, we have that

$$\frac{\partial u_m^0}{\partial x_n} \in W^{1,p}(\Omega) \subset C(\overline{\Omega}).$$

and hence $\dfrac{\partial u_m^0}{\partial x_n}$ is bounded. The estimate (4.46) turns then to

$$\int_{\Omega_\varepsilon} \left(\frac{\partial u_m^0}{\partial x_n} \right)^2 dx \leqslant C\varepsilon,$$

and the final estimate is

$$\|u^\varepsilon - u^{1\varepsilon}\|_{H^1(\Omega;\mathbb{R}^3)} \leqslant C\sqrt{\varepsilon}.$$

5 Conclusion

A three-dimensional mathematical model of an acoustic biosensor is developed. The model takes into account all the important structural components described by different physical laws. In particular, the surrounding liquid and the bristle-like layers at the solid-liquid interface are taken into account. The model provides time-periodic solutions only, which is conventional in acoustic applications because of linear behavior of materials at small deformations. The well-posedness of the model is rigorously established. In particular, it is mathematically proved that the elimination of nonzero eigenvalues of Helmholtz-like system can be achieved by posing special boundary conditions that physically express the contact of the sensor's side faces with a very viscose medium.

Since no assumptions about the polarization of the wave vector or the attenuation rate in the substrate are made, the model can easily be adjusted to simulate a wide range of acoustic devices surrounded by damping areas.

Along with pure theoretical investigation, the Ritz-Galerkin discretization of the problem is analyzed. The well-posedness of the discretized problem and the convergence of the Ritz-Galerkin solution to the exact one are established. Thus, the application of the finite element method is mathematically founded.

A numerical scheme based on a domain decomposition approach is developed. This is motivated by a large scale of the discretized problem. The proposed method is slower than the straightforward calculation, but it allows to handle a finer discretization without running out of the memory by distributing the discretized domain over several groups of nodes or clusters.

Finally, the results of 3D simulations are presented. The simulations have been carried out by the FE-program FeliCs developed at the Chair of Applied Mathematics of TU München. In order to increase the computational efficiency, the program has been enriched by parallel linear solvers making possible the parallel computation on high-performance clusters. Besides, in scope of the project VIOLA, we have developed a special version of

FeliCs that is able to compute on several heterogeneous clusters simultaneously. Using this program, we have carried out a number of simulations based on the domain decomposition scheme distributing the subdomains over the clusters located in different cities. However, the most of the simulations have been performed on the linux cluster of Leibniz-Rechenzentrum.

A semi-analytical method for the fast characterization of acoustic waves in multi-layered structures is developed. The method allows us to identify plane waves that can possibly exist in a given structure, compute their velocities and derive the dispersion relations. It can handle multi-layered structures composed of an arbitrary number of layers of different materials types. The method is efficient for the analysis of surface waves in very thick layers. In this case, layers are modeled as semi-infinite and only attenuating wave modes are considered. This allows to filter out negligible parasitic wave modes caused by the reflection from the bottom of a layer.

Beside elastic isotropic and anisotropic non-smart materials, the method is also able to treat piezoelectric materials and surrounding fluid media. For a more realistic modeling of piezoelectric layers, dielectric properties of non-piezoelectric adjacent layers or media (like gas or vacuum) can be taken into account. A special type of material is proposed to handle thin bristle-like interlayers arising in some sensors at the liquid-solid interface. The modeling of such layers is based on homogenization results presented in [28].

Another kind of homogenization is applied to handle composite materials consisting of a large number of periodically alternating thin sublayers. Here, the limiting equations are rigorously derived by the two-scale method for the general three-dimensional case. In the case of a layered structure, an explicit formula for the elasticity tensor of the homogenized material is obtained. This formula can be very useful for engineers and physicists investigating the elastic properties of multilayers. Furthermore, an error estimate for the important case where the right-hand side is from L^2 is established.

The method of dispersion relations is implemented in a computer program that provides a powerful modeling tool applicable in many areas. Concerning acoustic sensors, the program can be used for a fast analysis of many basic characteristics such as wavelength, velocity profiles, attenuation rate, sensitivity depending on crystal properties and thicknesses of the layers, and so on. A more accurate analysis of a sensor can be performed by the finite element method. In this case, the preliminary values obtained by the program can be used to adjust and optimize the finite element model. On the other hand, the program can be used for fast verification of finite element simulations.

In future, we plan to enrich the finite element model of the biosensor by taking into account not only electrical but also mechanical and geometrical properties of the electrodes. Another direction of the work is the homogenization of piezoelectric laminated structures. As in the case of elastic multilayers, we hope to derive here an explicit formula for the stress piezoelectric and dielectric tensors of the homogenized material.

Bibliography

[1] A. Abo-Zena. Dispersion function computations for unlimited frequency values. *Geophys. J. R. Astr. Soc.*, 58:91–105, 1979.

[2] Robert A. Adams. *Sobolev Spaces*. Pure and applied mathematics. New York: Academic Press, 1975.

[3] G. Allaire. Homogenization and two-scale convergence. *SIAM J. Math. Anal.*, 23(6):1482–1518, 1992.

[4] Bertram Alexander Auld. *Acoustic Fields and Waves in Solids*. Krieger Publishing Company, 1973.

[5] Ivo Babuška. Error-bounds for finite element method. *Numer. Math.*, 16:322–333, 1971.

[6] N. D. Botkin and V.L. Turova. Mathematical models of a biosensor. *Applied Mathematical Modelling*, 28:573–589, 2004.

[7] C Cai, G R Liu, and K Y Lam. A technique for modelling multiple piezoelectric layers. *Smart Mater. Struct.*, 10:689–694, 2001.

[8] C.Aschcraft and R. Grimes. Spooles: An object-oriented sparse matrix library. In *Proceedings of the 1999 SIAM Conference on Parallel Processing for Scientific Computing*, March 1999.

[9] M. Castaings and B. Hosten. Delta operator technique to improve the thomson-haskell method stability for propagation in multilayered anisotropic absorbing plates. *J. Acoust. Soc. Am.*, 95(4):1931–1941, 1994.

[10] Philippe G. Ciarlet. *Mathematical Elasticity, Volume I: Three-Dimensional Elasticity, Studies in Mathematics and its Applications 20*. North-Holland, Amsterdam, 1988.

[11] Doina Cioranescu and Patrizia Donato. *An Introduction to Homogenization*, volume 17 of *Oxford lecture series in mathematics and its applications*. Oxford University Press, 1999.

[12] A. Dall'aglio, V. De Cicco, D. Giachetti, and J.-P. Puel. Existence of bounded solutions for nonlinear elliptic equations in unbounded domains. *NoDEA : Nonlinear Differential Equations and Applications*, 11(4):431–450, December 2004.

[13] Supriyo Datta and Bill J. Hunsinger. Analysis of surface waves using orthogonal functions. *Journal of Applied Physics*, 49(2):475–479, 1978.

[14] H. Drobe, A. Leidl, M. Rost, and I. Ruge. Acoustic sensors based on surface-localized HPSWs for measurements in liquids. *Sensors and Actuators A: Physical*, 37-38:141–148, 1993.

[15] J. Du, G. L. Harding, J. A. Ogilvy, P. R. Dencher, and M. Lake. A study of love-wave acoustic sensors. *Sensors and Actuators*, A56:211–219, 1996.

[16] J.W. Dunkin. Computation of modal solutions in layered elastic media at high frequencies. *Bull. Seism. Soc. Am.*, 55:335–358, 1965.

[17] G. Endoh, K. Hashimoto, and M. Yamaguchi. Surface acoustic wave propagation characterization by finite-element method and spectral domain analysis. *Jpn. J. Appl. Phys.*, 34:2638–2641, 1995.

[18] Lawrence C. Evans. *Partial Differential Equations*. American Mathematical Society, 1998.

[19] F.Gilbert. Propagation of transient leaking modes in a stratified elastic wave guide. *Rev. Geophys.*, 2:123–153, 1964.

[20] N. Finger, G. Kovacs, J. Schöberl, and U. Langer. Accurate fem/bem-simulation of surface acoustic wave filters. *Ultrasonics, 2003 IEEE Symposium on*, 2:1680–1685, 2003.

[21] Andreas Gantner. *Mathematical Modeling and Numerical Simulation of Piezoelectrical Agitated Surface Acoustic Waves*. PhD thesis, Universität Augsburg, 2005.

[22] Andreas Gantner, Ronald H. W. Hoppe, Daniel Köster, Kunibert Siebert, and Achim Wixforth. Numerical simulation of piezoelectrically agitated surface acoustic waves on microfluidic biochips. *Comput. Vis. Sci.*, 10(3):145–161, 2007.

[23] G.Kovacs and M.Venema. Theoretical comparison of sensitivities of acoustic shear wave modes for biochemical sensing in liquids. *Appl Phys Lett*, 61:639–641, 1992.

[24] M. Gunzburger and S. Manservisi. Flow matching by shape design for the navier-stokes system. *International series of numerical mathematics*, 139:279–289, 2001.

[25] R. Hassdorf, M. Grimsditch, W. Felsch, and O. Schulte. Shear-elasticity anomaly in ce/fe multilayers. *Physical Review B.*, 56(10):5814–5821, 1997.

[26] M. Hofer, N. Finger, G.Kovacs, J. Schoberl, S. Zaglmayr, U. Langer, and R. Lerch. Finite-element simulation of wave propagation in periodic piezoelectric saw structures. *Ultrasonics, Ferroelectrics and Frequency Control, IEEE Transactions on*, 53:1192 – 1201, 2006.

[27] M. Hofer, M. Jungwirth, R. Lerch, and R. Weigel. Accurate and efficient modeling of saw structures. *Frequenz*, 55(1-2):64–72, 2001.

[28] K.-H. Hoffmann, N. D. Botkin, and V. N. Starovoitov. Homogenization of interfaces between rapidly oscillating fine elastic structures and fluids. *SIAM J. Appl. Math.*, 65(3):983–1005, 2005.

[29] Lars Hörmander. *The analysis of linear partial differential operators*, volume I. Springer, 1990.

[30] David Bau III and Lloyd N. Trefethen. *Numerical linear algebra.* SIAM: Society for Industrial and Applied Mathematics, 1997.

[31] B. Jakobi and M.J. Vellekoop. Properties of love waves: applications in sensors. *Smart Mater. Struct.*, 6:668–769, 1997.

[32] B. Jakobi and M.J. Vellekoop. Viscosity sensing using a love-wave device. *Sensors and Actuators*, A68:275–281, 1998.

[33] V.V. Jikov, S.M. Kozlov, and O.A. Oleinik. *Homogenization of Differential Operators and Integral Functionals*. Springer-Verlag, 1994.

[34] Yoonkee Kim and William D. Hunt. Acoustic fields and velocities for surface-acoustic-wave propagation in multilayered structures: An extension of the laguerre polynomial approach. *Journal of Applied Physics*, 68(10):4993–4997, 1990.

[35] Yoonkee Kim and William D. Hunt. An analysis of surface acoustic wave propagation in a piezoelectric film over a gaas/algaas heterostructure. *Journal of Applied Physics*, 71(5):2136–2142, 1992.

[36] L. Knopoff. A matrix method for elastic wave problems. *Bull. Seism. Soc. Am.*, 54:431–438, 1964.

[37] Daniel Köster. *Numerical Simulation of Acoustic Streaming on SAW-driven Biochips.* PhD thesis, Augsburg, 2006.

[38] Daniel Köster. Numerical simulation of acoustic streaming on surface acoustic wave-driven biochips. *SIAM J. Sci. Comput.*, 29, no. 6:2352–2380, 2007.

[39] G. Kovacs, M.J. Vellekoop, R. Haueis, G.W. Lubking, and A. Venema. A love wave sensor for (bio)chemical sensing in liquids. *Sensors and Actuators*, A43:38–43, 1994.

[40] L.D. Landau and E.M. Lifschitz. *Elastizitätstheorie.* Akademie Verlag, Berlin, 1975.

[41] L.D. Landau and E.M. Lifschitz. *Hydrodynamik.* Akademie Verlag, Berlin, 1975.

[42] J. E. Lefebvre, V. Zhang, J. Gazalet, and T. Gryba. Conceptual advantages and limitations of the laguerre polynomial approach to analyze surface acoustic waves in semi-infinite substrates and multilayered structures. *Journal of Applied Physics*, 83(1):28–34, 1998.

[43] J. E. Lefebvre, V. Zhang, J. Gazalet, and T. Gryba. Legendre polynomial approach for modeling free-ultrasonic waves in multilayered plates. *Journal of Applied Physics*, 85(7):3419–3427, 1999.

[44] R. Lerch. Simulation of piezoelectric devices by two- and three-dimensional finite elements. *Ultrasonics, Ferroelectrics and Frequency Control, IEEE Transactions on*, 37(3):233–247, 1990.

[45] D. Lévesque and L. Piché. A robust transfer matrix formulation for the ultrasonic response of multilayered absorbing media. *J. Acoust. Soc. Am.*, 92:452–467, 1992.

[46] J.-L. Lions. *Quelques Methodes de Resolution des Problemes aux Limites non Lineaires.* Dunod Gauthier-Villars, Paris, 1969.

[47] A.E.H. Love. *Some problems of geodynamics.* Cambridge University Press, London, 1911.

[48] Michael J. S. Lowe. *Plate waves for the NDT of diffusion bonded titanium', Mechanical Engineering.* PhD thesis, Imperial College London, 1993.

[49] Michael J.S. Lowe. Matrix techniques for modeling ultrasonic waves in multilayered-media. *Ultrasonics, Ferroelectrics and Frequency Control, IEEE Transactions on*, 42(4):525–542, 1995.

[50] Dag Lukkassen, Gabriel Ngutseng, and Peter Wall. Two-scale convergence. *International Journal of Pure and Applied Mathematics*, 2(1):35–86, 2002.

[51] Vladimir Maz'ya. *Sobolev Spaces.* Springer-Verlag, 1985.

[52] R. Miller. *Acoustic Interactions with Submerged Elastic Structures*, chapter 10, pages 278–328. World Scientific, 1996.

[53] J. Nečas. Sur une méthode pour résoudre les équations aux dérivées partielles du type elliptique, voisine de la variationnelle. *Ann. Sc. Norm. Super. Pisa, Sci. Fis. Mat., III. Ser.*, 16:305–326, 1962.

[54] Gabriel Ngutseng. A general convergence result for a functional related to the theory of homogenization. *SIAM J. Math. Anal.*, 20:608–629, 1989.

[55] Otto Nikodym. Sur une classe de fonctions considérées le problème de dirichlet. *Fundamenta Mathematicae*, 21:129–150, 1933.

[56] O.A. Oleinik, A.S. Shamaev, and G.A. Yosifian. *Mathematical Problems in Elasticity and Homogenization.* Elsevier Science Ltd, 1992.

[57] B. Pavlakovic, M.J.S. Lowe, D.N. Alleyne, and P. Cawley. Disperse: a general purpose program for creating dispersion curves. *Review of Progress in Quantitative NDE*, 16:185–192, 1997.

[58] R.A. Phinney. Propagation of leaking interface waves. *Bull. Seism. Soc. Am.*, 51:527–555, 1961.

[59] M. Åberg and P. Gudmundson. The usage of standard finite element codes for computation of dispersion relations in materials with periodic microstructure. *J. Acoust. Soc. Am.*, 102(4):2007–2013, 1997.

[60] Daniel Royer and Eugène Dieulesaint. *Elastic Waves in Solids: Free and guided propagation.* Springer-Verlag Berlin-Heidelberg-New York, 2000.

[61] O. Schenk and K. Gärtner. Solving unsymmetric sparse systems of linear equations with pardiso. *Journal of Future Generation Computer Systems*, 20(3):475–487, 2004.

[62] H. Schmidt and F.B. Jensen. Efficient numerical solution technique for wave propagation in horizontally stratified environments. *Comp. & Maths. with Appls.*, 11:699–715, 1985.

[63] H. Schmidt and G. Tango. Efficient global matrix approach to the computation of synthetic seismograms. *Geophys. J. R. Astr. Soc.*, 84:331–359, 1986.

[64] F.A. Schwab. Surface-wave dispersion computations: Knopoff's method. *Bull. Seism. Soc. Am.*, 60:1491–1520, 1970.

[65] F.A. Schwab and L. Knopoff. Surface waves on multilayered anelastic media. *Bull. Seism. Soc. Am.*, 61:893–912, 1971.

[66] F.A. Schwab and L. Knopoff. Fast surface wave and free mode computations. *Methods in computational physics*, IX:87–180, 1972.

[67] Guido Stampacchia. Equations elliptiques du second ordre à coefficients discontinus. *Séminaire Jean Leray*, 3:1–77, 1963-1964.

[68] Luc Tartar. Estimations des coefficients homogénéisés. *Lectures Notes in Mathematics*, 704:364–373, 1977.

[69] Luc Tartar. Quelques remarques sur l'homogeneisation. In *Functional Analysis and Numerical Analysis, (Japan-France Seminar, Japanese Society for the Promotion of Science)*, pages 469–482, 1978.

[70] William T. Thomson. Transmission of elastic waves through a stratified solid medium. *J. Appl. Phys.*, 21:89–93, 1950.

[71] E.N. Thrower. The computation of dispersion of elastic waves in layered media. *J. Sound Vib.*, 2:210–226, 1965.

[72] Pascal Tierce, Koffi Anifrani, Jean Noel Decarpigny, and B. Hamonic. Piezoelectric transducer analysis using the transfer matrix method and a new versatile computer code. *The Journal of the Acoustical Society of America*, 64(S1):S102–S102, 1988.

[73] T.H. Watson. A real frequency, complex wave-number analysis of leaking modes. *Bull. Seism. Soc. Am.*, 62:369–541, 1972.

[74] W.Hackbusch. *Elliptic Differential Equations*. Springer Verlag, 1992.

[75] Jiří Zelenka. *Piezoelectric Resonators and their Applications (Studies in Electrical and Electronic Engineering)*. Elsevier Science Publishers B. V., 1986.

MIX
Papier aus verantwortungsvollen Quellen
Paper from responsible sources
FSC® C105338

Printed by Books on Demand GmbH, Norderstedt / Germany